Multi-objective Management in Freight Logistics

Massimiliano Caramia • Paolo Dell'Olmo

Multi-objective Management in Freight Logistics

Increasing Capacity, Service Level and Safety with Optimization Algorithms

Springer

Massimiliano Caramia, PhD
Università di Roma "Tor Vergata"
Dipartimento di Ingegneria dell'Impresa
Via del Politecnico, 1
00133 Roma
Italy

Paolo Dell'Olmo, PhD
Università di Roma "La Sapienza"
Dipartimento di Statistica, Probabilità
e Statistiche Applicate
Piazzale Aldo Moro, 5
00185 Roma
Italy

ISBN 978-1-84800-381-1 e-ISBN 978-1-84800-382-8

DOI 10.1007/978-1-84800-382-8

British Library Cataloguing in Publication Data
Caramia, Massimiliano
Multi-objective management in freight logistics :
increasing capacity, service level and safety with
optimization algorithms
1. Freight and freightage - Mathematical models 2. Freight
and freightage - Management 3. Business logistics
I. Title II. Dell'Olmo, Paolo, 1958-
388'.044'015181
ISBN-13: 9781848003811

Library of Congress Control Number: 2008935034

Cover design: eStudio Calamar S.L., Girona, Spain

Printed on acid-free paper

9 8 7 6 5 4 3 2 1

springer.com

Preface

The content of this book is motivated by the recent changes in global markets and the availability of new transportation services. Indeed, the complexity of current supply chains suggests to decision makers in logistics to work with a set of efficient (Pareto-optimal) solutions, mainly to capture different economical aspects that, in general, one optimal solution related to a single objective function is not able to capture entirely. Motivated by these reasons, we study freight transportation systems with a specific focus on multi-objective modelling. The goal is to provide decision makers with new methods and tools to implement multi-objective optimization models in logistics. The book combines theoretical aspects with applications, showing the advantages and the drawbacks of adopting scalarization techniques, and when it is worthwhile to reduce the problem to a goal-programming one. Also, we show applications where more than one decision maker evaluates the effectiveness of the logistic system and thus a multi-level programming is sought to attain meaningful solutions. After presenting the general working framework, we analyze logistic issues in a maritime terminal. Next, we study multi-objective route planning, relying on the application of hazardous material transportation. Then, we examine freight distribution on a smaller scale, as for the case of goods distribution in metropolitan areas. Finally, we present a human-workforce problem arising in logistic platforms. The general approach followed in the text is that of presenting mathematics, algorithms and the related experimentations for each problem.

Rome,
May 2008

Massimiliano Caramia
Paolo Dell'Olmo

Acknowledgements

The authors wish to thank Eugenia Civico, Isabella Lari, Alessandro Massacci and Maria Grazia Mecoli for their contribution in the implementation of the models in Chapter 3; Pasquale Carotenuto, Monica Gentili and Andrea Scozzari for their contribution to the hazmat project to which Chapter 4 refers; Monica Gentili and Pitu Mirchandani as co-authors of previous work related to Chapter 6.

Contents

List of Figures

List of Tables

Chapter 1
Introduction

Abstract In this chapter, we introduce freight distribution logistic, discussing some statistics about future trends in this area. Basically, it appears that, even though there has been a slight increase in the use of rail and water transportation modes, there is room to obtain a more efficient use of the road mode, mainly not to increase air pollution (fossil-fuel combustion represents about 80% of the factors that jeopardize air quality). In order to be able to reach an equilibrium among different transportation modes, the entire supply chain has to be studied to install the appropriate service capacity and to define effective operational procedures to optimize the system performance.

1.1 Freight Distribution Logistic

The way the governments and the economic world are now looking at transportation problems in general and at distribution logistic in particular has changed from past years. Europe, in particular, has adopted the so-called European Sustainable Development Strategy (SDS) in the European Council held in Gothenburg in June 2001. That was the opportunity to set out a coherent approach to sustainable development renewed in June 2006 to reaffirm the aim of continuous improvement of quality of life and economic growth through the efficient use of resources, and promoting the ecological and the social innovation potentials of the economy. Recently, in December 2007, the European Council insisted on the need to give priority to implementation measures. Paragraph 56 of Commission Progress Report of 22 October 2007 reads "... The EU must continue to work to move towards more sustainable transport and environmentally friendly transport modes. The Commission is invited to present a roadmap together with its next Progress Report in June 2009 on the SDS setting out the remaining actions to be implemented with highest priority."

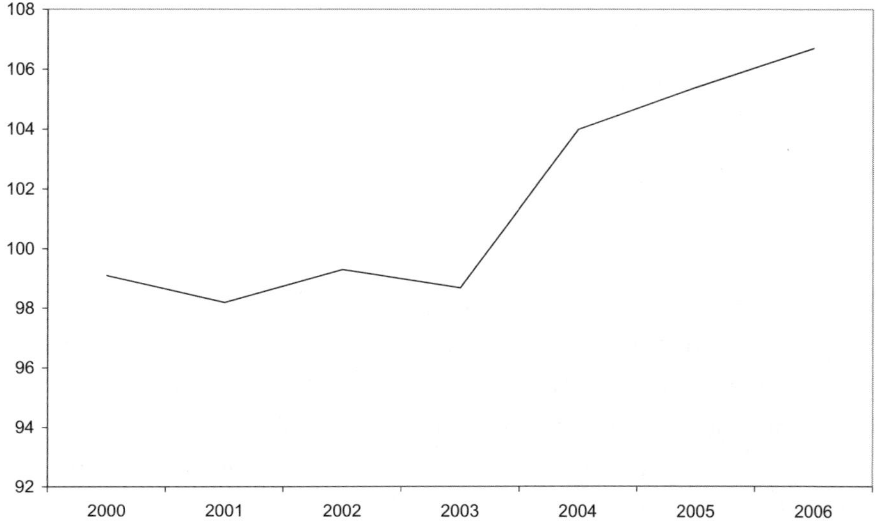

Fig. 1.1 Transported goods in Europe in the last years (source Eurostat, www.epp.eurostat.ec.europa.eu). The *y*-axis reports the ratio between tonne-kilometers and gross domestic product in Euros indexed on 1995. The index includes transport by road, rail and inland waterways

Focusing on freight transportation, which has a significant impact on the social life of inhabitants especially in terms of air pollution and land usage, we show the current trend of the volume of transported goods in Europe in Fig. 1.1.

Nowadays, the European transport system cannot be said to be sustainable, and in many respects it is moving away from sustainability. The European Environment Agency (EEA) highlights, in particular, the sector's growing CO_2 emissions that threaten the Kyoto protocol target (see Fig. 1.2). The chart reports the carbon dioxide (CO_2) values (Mio. t CO_2) that is the most relevant greenhouse gas (GHG); indeed, it is responsible, for about 80%, of the global-warming potential due to anthropogenic GHG emissions covered by the Kyoto Protocol. Fossil-fuel burning is the main source of CO_2 production.

The EEA also points to additional efforts that are needed to reach existing air-quality targets, risks to population and so on. In a recent study of Raupach et al.

(2007) about "Global and regional drivers of accelerating CO_2 emissions", the authors reported that in the years 2000–2004 the percentage of factors contributing to CO_2 are as follows:

1. Solid fuels (e.g., coal): 35%
2. Liquid fuels (e.g., gasoline): 36%
3. Gaseous fuels (e.g., natural gas): 20%
4. Flaring gas industrially and at wells: <1%
5. Cement production: 3%
6. Non-fuel hydrocarbons: <1%
7. The "international bunkers" of shipping and air transport not included in national inventories: 4%.

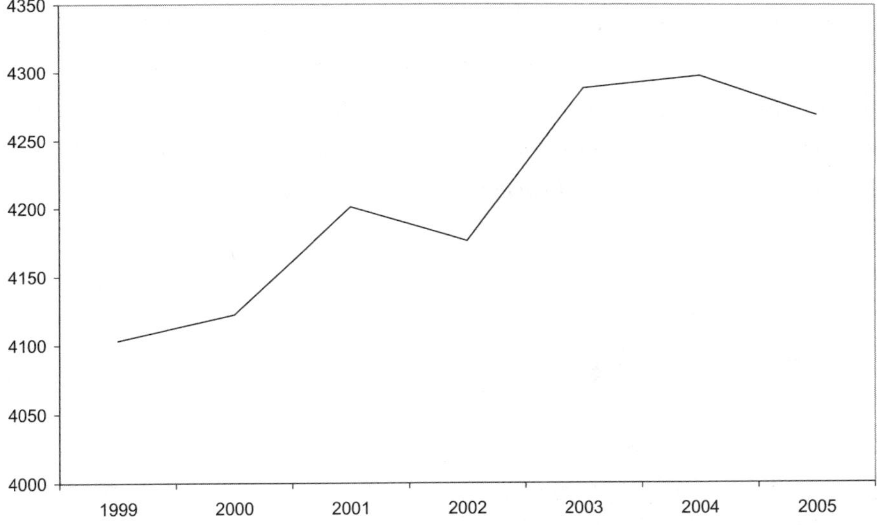

Fig. 1.2 Pollution trend with time in Europe (source Eurostat, www.epp.eurostat.ec.europa.eu). The *y*-axis reports the carbon dioxide (CO_2) values (Mio. t CO_2)

Because of the underlying complexity of the mentioned problems a sensitive benefit to policy and decision makers working in this arena may be taken from the adoption of formal decisions methods. Even though the literature proposes many effective decision-support models and systems that can be used in this direction, we believe that other research has to be done. In particular, the complexity of the practical problems suggest that different criteria (e.g., economical, environmental, social) have to be taken into account, shedding light on the goal of this book. This is mainly that of showing how multiple objectives can be managed in freight logistic and transportation; therefore, most of the models we propose are multi-criteria decision ones. Our description of distribution networks has a twofold characteristic: on the one hand, it is a top-down view of the big picture, and, on the other hand, it is also oriented to point out which collection of data is required to feed the formal models. Moreover, one of the main objectives is to make the strategy of integration among different modes more operational. This arises from a simple observation on the distribution of volumes among different modes, as reported in Fig. 1.3. It is notable that the considerable growth in transport has been almost entirely realized by road transport. In 2000 it represented 75% of the tonne-kilometers performed in the European countries. Even though the railway goods transport increased during recent years, this growth has not been at the same pace as road transport. The conclusion is that the number of tonne-kilometers by road is much greater than the tonne-kilometers performed by rail. We note that in 1995 the million tonne-kilometers of goods transported by rail was about 220,000 while in 2005 it was about 262,000. Inland waterway transport progressed by only 17% in nearly three decades. We note that in 1995 the index related to waterways transportation was 114,394.

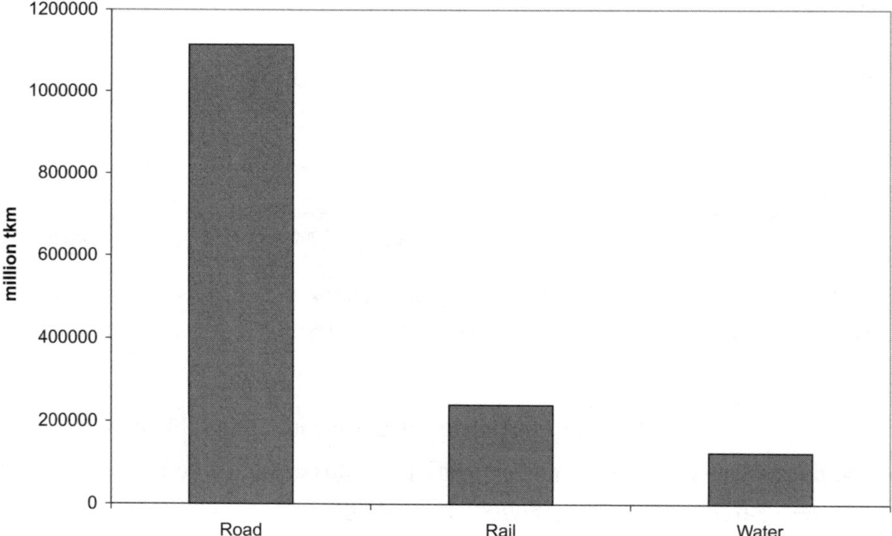

Fig. 1.3 Data refer to Europe, year 2000 (source Eurostat, www.epp.eurostat.ec.europa.eu). Data account for inland transport only

As far as different modes are concerned we may consider an example of a supply chain like the one depicted in Fig. 1.4.

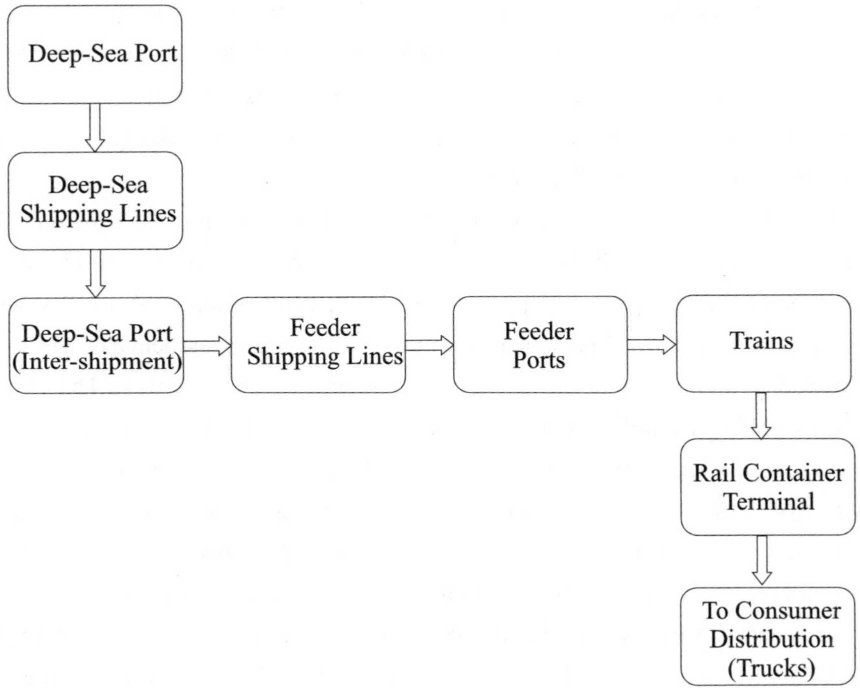

Fig. 1.4 Supply-chain representation of different transportation mode flow

In this flow chart, the chain starts with a deep-sea port from where deep-sea shipping lines deliver containers to an inter-shipment port. From here smaller vessels can reach other harbors from where we may assume that trains deliver containers to railway container terminals. From these terminals, goods are delivered to customers in different ways, but most likely by means of trucks. Truck-route definition problems cover a relevant role in freight logistics and have stimulated the research of a large amount of optimization problems, and among them the *vehicle routing problem* (VRP) and all its variants are without doubt the most studied ones.

Freight logistic distribution problems can be modelled as combinatorial optimization problems on transportation networks. A transportation network is represented by means of a graph $G = (V, A)$, where the collection of arcs A represents the main viable ways, e.g., motorways, seaways and railways, and the node set V

represents relevant intersections among these ways. Arcs can be either directed or undirected based on the network type, e.g., there are problems in which it can be assumed that viable ways can be traversed in both directions with the same time and/or cost, and, therefore, it is redundant to specify the orientation of the arc. Clients as well as facilities are located in the nodes of the networks. Note that sometimes transportation networks are modelled by means of hypergraphs (e.g., see Nguyen et al., 1998), but this in general happens in public transportation network representations that are beyond the scope of this book.

VRP is seen as one of the most critical elements in managing the supply chain (see, e.g., Laporte et al., 2000, Toth and Vigo, 2002, Cordeau et al., 2002). The basic VRP can be stated as follows. Given are one depot, a fleet of identical vehicles with given capacities, a set of customers with given locations, given demands for each commodity, with all the distances measured in terms of length or time. The objectives of VRP can be different, e.g., one could be interested in finding the minimum cost (length) route or the minimum number of vehicles, such that each customer is serviced exactly once, every route originates and ends in the depot, the capacity constraints are respected. VRP models are used not only to give routes and drivers directions, but also to sequence stops on routes and schedule stops for pick-up and delivery tasks. Variations of the standard VRP include time windows (TWVRP) (see, e.g., Madsen, 2005) where a limited time interval is given for each location to be visited and the dynamic version (DVRP) that models the practical issue that new requests are known only when some of the routes are already planned (see, e.g., Madsen et al., 2007).

As vehicles are capacitated, and in general other resources have limited capacity, sometimes VRP is solved in two steps: in the first step tasks (goods, trips, containers, customers) are assigned to capacitated vehicles (or other kinds of resources), and in the second phase, starting from the assignment of the first phase, the route-definition problem is solved. The advantage of this approach is that of reducing the practical problem complexity splitting the original problem into two subproblems. The drawback is that often a heuristic algorithm is needed to retrieve feasibility or to improve the solution quality, after the two phases have been carried out. This approach is often used in the so-called *truck and trailer routing problem* (TTRP). Indeed, a general assumption on VRP is that customers can be reached by the vehicles without exception, i.e., the size of a truck and/or the location of the customer is not relevant. In practice, this could not be true; indeed, vehicles can have trailers and sometimes a truck with its trailer cannot reach a subset of the customer locations, in which situation the trailer has to be uncoupled from the *complete vehicle* (i.e., the truck plus the trailer) in ad-hoc parking areas, and after visiting these customers the trailer is coupled again to the *pure truck* forming again a complete vehicle that continues its tour. For the TTRP problem the reader is referred to, e.g., Chao (2002) and Scheuerer (2006).

Referring again to the fleet of vehicles, one can also distinguish between a homogeneous and a heterogeneous fleet. This generates a further variant of VRP denoted as the *heterogeneous vehicle routing problem* (HVRP) where vehicles have different capacities (see, e.g., Baldacci, 2007). However, even if the HVRP considers transport means with different capacities, means are considered homogeneous with respect to the transportation mode, that is all the vehicles belong to the road mode. Therefore, we can further distinguish two kinds of freight distribution, one denoted as *less than truckload* (LTL) transportation and the other denoted *full truckload* (FTL) transportation. For a comprehensive study on LTL and FTL transportation the reader can see the 2002 survey of Crainic on "Long Haul Freight Transportation".

LTL is a service offered by many freight and trucking companies for businesses that only need a small shipment of goods delivered. In contrast, a full truckload or large shipment uses all available space in a tractor trailer. For a recent study on "Vehicle routing and scheduling with full truckloads" the reader is referred to the 2003 paper of Arunapuram et al.

A less than truckload shipment is delivered with other shipments and, in general, is not directly processed to a destination as in FTL transportation.

LTL shipments are arranged so that the driver picks up the shipment along a short route and brings it back to a logistic platform, where it is processed in order to be transferred to another truck. The latter brings the shipment, along with other small shipments, to another city terminal. The LTL shipment is then moved from truck to truck until it reaches its final destination.

LTL, as compared to FTL, has one main advantage and one main drawback: the former is economical since the cost of shipping less than a truckload is relatively inexpensive, and in particular is less than the cost of a FTL transportation. On the other hand, however, a LTL shipment can have longer processing times to reach to destination than a FTL shipment, since it does not follow a direct route from its origin to its final client.

We need to say that LTL is not a parcel-carrier service. It can be located in between the latter and FTL shipment. It uses, in general, trucks with trailers, as in FTL transportation, but behaves like shipments handled by parcel carriers. As a parcel service that tends to handle large packages, companies like UPS, DHL, FedEx, are good examples of how a LTL shipment involves repeated transfers. A driver of one of these companies picks up a shipment along his/her route, which is brought back to the terminal at the end of his shift. The shipment is then loaded onto an overnight truck and transferred again through a daily route. This process is repeated until the shipment reaches its final destination.

These distinctions are important since we have to specify two kinds of problems: those in which shipments are performed by means of, e.g., road mode only, and those processed by means of multiple transportation modes. In the first type of problems we have to make a further distinction, i.e., road mode transportation performed by using a single transport mean and road mode transportation made up of more than one transport mean. In the latter case, we speak of inter-modality. Referring to this latter aspect, Caramia and Guerriero (2008) studied a multi-objective long-haul freight transportation problem, where travel time and route cost are to be minimized together with the maximization of a transportation mean sharing index, related to the capability of the transportation system of generating economy-scale solutions. In terms of constraints, besides vehicle capacity and time windows, transportation jobs have to obey additional constraints related to mandatory nodes (e.g., logistic platform nearest to the origin or the destination) and forbidden nodes (e.g., logistic platforms not compatible with the operations required). We refer the reader to the book of Ghiani et al. (2004) for problems and models on this kind of problems.

Multi-objective optimization is quite natural in vehicle routing: indeed, there could be solutions that, e.g., minimize the number of vehicles with long length routes, and other solutions that minimize the route lengths with a large number of vehicles. If one considers that minimizing the number of vehicles directly affects the vehicle costs and the labor costs, while minimizing route length is directly related to fuel and time costs, one clearly realizes that prioritizing objectives to deal with a single objective approach is very difficult.

In the literature, authors have investigated the problem with either two or three objective functions, and with different optimization techniques. The minimization of the number of routes and of the travel costs has been considered in, e.g., Ombuki et al. (2006) where the authors proposed a genetic algorithm approach to cope with this problem. Gambardella et al. (1999) studied the problem with the minimization of the number of vehicles and the total costs with a hierarchical approach, in which they designed two ant colonies each one dedicated to the optimization of an objective function. Murata and Itai (2005) considered the minimization of the number of vehicles and of the maximum routing time among the vehicles, using evolutionary multi-criterion optimization algorithms. Liu et al. (2006) studied the problem with three objective functions, i.e., the total distance travelled by the vehicles, the balance workloads and the balance delivery times among the dispatch vehicles. They transformed the starting multi-objective program into a goal programming one, and proposed a heuristic based on one-point movement, two-point exchange, and intra-route one-exchange local searches. Seo and Choi (1998) presented a genetic algorithm based search technique to find alternative paths between origin–destination pairs. The method can provide multiple alternatives that are nearly optimal, and is able to reduce similarities among the paths.

Before closing this section we present the outline of the book. Chapter 2 deals with multi-objective optimization. Here, we will discuss the main techniques to cope with such problems.

In Chap. 3, we will look at freight distribution problems inherent to a maritime terminal, that represents the origin of the road and the rail shipments. This analysis also has the objective of introducing how simulation tools can be used to set capacity and service levels.

In Chaps. 4 and 5, we will take into account two problems, i.e., hazardous-material transportation, and district logistic, highlighting the multi-objective nature embedded in these applications. The choice of these two problems stems from the relevant impact on safety that hazardous material transportation accidents can produce on the neighboring population, and on air pollution in cities worldwide, respectively.

Chapter 6 discusses a staff-scheduling problem, with particular focus on logistic platforms, discussing also some multi-modal and inter-modal issues inherent to this topic.

Chapter 2
Multi-objective Optimization

Abstract In this chapter, we introduce multi-objective optimization, and recall some of the most relevant research articles that have appeared in the international literature related to these topics. The presented state-of-the-art does not have the purpose of being exhaustive; it aims to drive the reader to the main problems and the approaches to solve them.

2.1 Multi-objective Management

The choice of a route at a planning level can be done taking into account time, length, but also parking or maintenance facilities. As far as advisory or, more in general, automation procedures to support this choice are concerned, the available tools are basically based on the "shortest-path problem". Indeed, the problem to find the single-objective shortest path from an origin to a destination in a network is one of the most classical optimization problems in transportation and logistic, and has deserved a great deal of attention from researchers worldwide. However, the need to face real applications renders the hypothesis of a single-objective function to be optimized subject to a set of constraints no longer suitable, and the introduction of a multi-objective optimization framework allows one to manage more information. Indeed, if for instance we consider the problem to route hazardous materials in a road network (see, e.g., Erkut et al., 2007), defining a single-objective function problem will involve, separately, the distance, the risk for the population, and the transportation costs. If we regard the problem from different points of view, i.e., in terms of social needs for a safe transshipment, or in terms of economic issues or pollution reduction, it is clear that a model that considers simultaneously two or more such objectives could produce solutions with a higher level of equity. In the following, we will discuss multi-objective optimization and related solution techniques.

2.2 Multi-objective Optimization and Pareto-optimal Solutions

A basic single-objective optimization problem can be formulated as follows:

$$\min f(x)$$
$$x \in S,$$

where f is a scalar function and S is the (implicit) set of constraints that can be defined as

$$S = \{x \in R^m : h(x) = 0, g(x) \geq 0\}.$$

Multi-objective optimization can be described in mathematical terms as follows:

$$\min [f_1(x), f_2(x), \ldots, f_n(x)]$$
$$x \in S,$$

where $n > 1$ and S is the set of constraints defined above. The space in which the objective vector belongs is called the *objective space*, and the image of the feasible set under F is called the *attained set*. Such a set will be denoted in the following with

$$C = \{y \in R^n : y = f(x), x \in S\}.$$

The scalar concept of “optimality” does not apply directly in the multi-objective setting. Here the notion of Pareto optimality has to be introduced. Essentially, a vector $x^* \in S$ is said to be Pareto optimal for a multi-objective problem if all other vectors $x \in S$ have a higher value for at least one of the objective functions f_i, with $i = 1, \ldots, n$, or have the same value for all the objective functions. Formally speaking, we have the following definitions:

- A point x^* is said to be a *weak* Pareto optimum or a *weak* efficient solution for the multi-objective problem if and only if there is no $x \in S$ such that $f_i(x) < f_i(x^*)$ for all $i \in \{1, \ldots, n\}$.

- A point x^* is said to be a *strict* Pareto optimum or a *strict* efficient solution for the multi-objective problem if and only if there is no $x \in S$ such that $f_i(x) \leq f_i(x^*)$ for all $i \in \{1,\ldots,n\}$, with at least one strict inequality.

We can also speak of locally Pareto-optimal points, for which the definition is the same as above, except that we restrict attention to a feasible neighborhood of x^*. In other words, if $B(x^*,\varepsilon)$ is a ball of radius $\varepsilon > 0$ around point x^*, we require that for some $\varepsilon > 0$, there is no $x \in S \cap B(x^*,\varepsilon)$ such that $f_i(x) \leq f_i(x^*)$ for all $i \in \{1,\ldots,n\}$, with at least one strict inequality.

The image of the efficient set, i.e., the image of all the efficient solutions, is called Pareto front or Pareto curve or surface. The shape of the Pareto surface indicates the nature of the trade-off between the different objective functions. An example of a Pareto curve is reported in Fig. 2.1, where all the points between $(f_2(\hat{x}), f_1(\hat{x}))$ and $(f_2(\tilde{x}), f_1(\tilde{x}))$ define the Pareto front. These points are called non-inferior or non-dominated points.

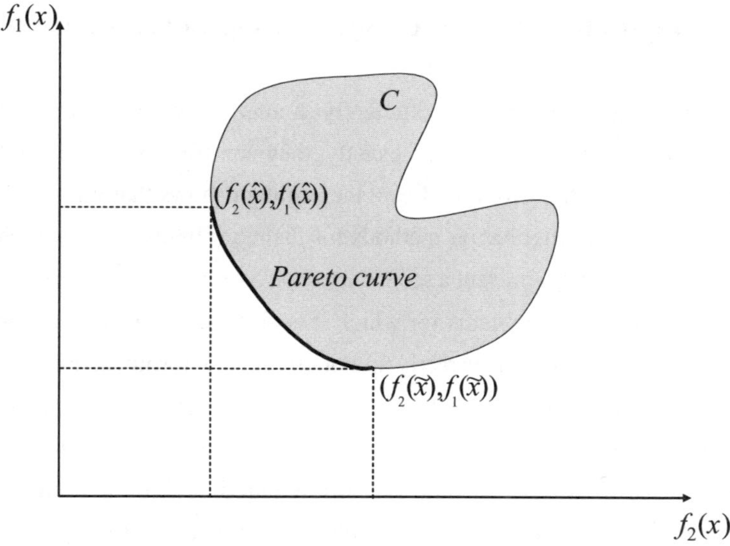

Fig. 2.1 Example of a Pareto curve

An example of weak and strict Pareto optima is shown in Fig. 2.2: points p_1 and p_5 are weak Pareto optima; points p_2, p_3 and p_4 are strict Pareto optima.

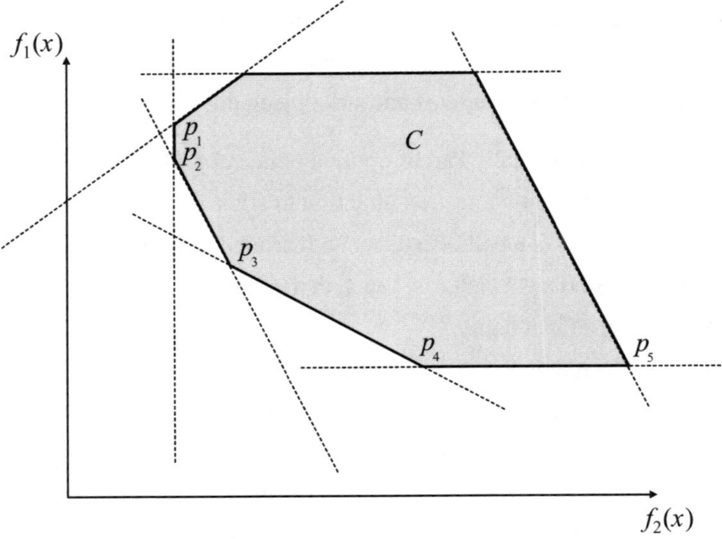

Fig. 2.2 Example of weak and strict Pareto optima

2.3 Techniques to Solve Multi-objective Optimization Problems

Pareto curves cannot be computed efficiently in many cases. Even if it is theoretically possible to find all these points exactly, they are often of exponential size; a straightforward reduction from the knapsack problem shows that they are NP-hard to compute. Thus, approximation methods for them are frequently used. However, approximation does not represent a secondary choice for the decision maker. Indeed, there are many real-life problems for which it is quite hard for the decision maker to have all the information to correctly and/or completely formulate them; the decision maker tends to learn more as soon as some preliminary solutions are available. Therefore, in such situations, having some approximated solutions can help, on the one hand, to see if an exact method is really required, and, on the other hand, to exploit such a solution to improve the problem formulation (Ruzica and Wiecek, 2005).

Approximating methods can have different goals: representing the solution set when the latter is numerically available (for convex multi-objective problems); approximating the solution set when some but not all the Pareto curve is numerically available (see non-linear multi-objective problems); approximating the solution set when the whole efficient set is not numerically available (for discrete multi-objective problems).

A comprehensive survey of the methods presented in the literature in the last 33 years, from 1975, is that of Ruzica and Wiecek (2005). The survey analyzes separately the cases of two objective functions, and the case with a number of objective functions strictly greater than two. More than 50 references on the topic have been reported. Another interesting survey on these techniques related to multiple objective integer programming can be found in the book of Ehrgott (2005) and the paper of Erghott (2006), where he discusses different scalarization techniques. We will give details of the latter survey later in this chapter, when we move to integer linear programming formulations. Also, T'Kindt and Billaut (2005) in their book on "Multicriteria scheduling", dedicated a part of their manuscript (Chap. 3) to multi-objective optimization approaches.

In the following, we will start revising, following the same lines of Erghott (2006), these scalarization techniques for general continuous multi-objective optimization problems.

2.3.1 The Scalarization Technique

A multi-objective problem is often solved by combining its multiple objectives into one single-objective scalar function. This approach is in general known as the *weighted-sum* or *scalarization* method. In more detail, the weighted-sum method minimizes a positively weighted convex sum of the objectives, that is,

$$\min \sum_{i=1}^{n} \gamma_i \cdot f_i(x)$$
$$\sum_{i=1}^{n} \gamma_i = 1$$
$$\gamma_i > 0,\ i = 1, \ldots, n$$
$$x \in S,$$

that represents a new optimization problem with a unique objective function. We denote the above minimization problem with $P_s(\gamma)$.

It can be proved that the minimizer of this single-objective function $P(\gamma)$ is an efficient solution for the original multi-objective problem, i.e., its image belongs to the Pareto curve. In particular, we can say that if the γ weight vector is strictly greater than zero (as reported in $P(\gamma)$), then the minimizer is a *strict* Pareto optimum, while in the case of at least one $\gamma_i = 0$, i.e.,

$$\min \sum_{i=1}^{n} \gamma_i \cdot f_i(x)$$
$$\sum_{i=1}^{n} \gamma_i = 1$$
$$\gamma_i \geq 0, \ i = 1, \ldots, n$$
$$x \in S,$$

it is a *weak* Pareto optimum. Let us denote the latter problem with $P_w(\gamma)$.

There is not an *a-priori* correspondence between a weight vector and a solution vector; it is up to the decision maker to choose appropriate weights, noting that weighting coefficients do not necessarily correspond directly to the relative importance of the objective functions. Furthermore, as we noted before, besides the fact that the decision maker cannot be aware of which weights are the most appropriate to retrieve a satisfactorily solution, he/she does not know in general how to change weights to consistently change the solution. This means also that it is not easy to develop heuristic algorithms that, starting from certain weights, are able to define iteratively weight vectors to reach a certain portion of the Pareto curve.

Since setting a weight vector conducts to only one point on the Pareto curve, performing several optimizations with different weight values can produce a considerable computational burden; therefore, the decision maker needs to choose which different weight combinations have to be considered to reproduce a representative part of the Pareto front.

Besides this possibly huge computation time, the scalarization method has two technical shortcomings, as explained in the following.

- The relationship between the objective function weights and the Pareto curve is such that a uniform spread of weight parameters, in general, does not produce a uniform spread of points on the Pareto curve. What can be observed about this fact is that all the points are grouped in certain parts of the Pareto front, while some (possibly significative) portions of the trade-off curve have not been produced.

- Non-convex parts of the Pareto set cannot be reached by minimizing convex combinations of the objective functions. An example can be made showing a geometrical interpretation of the weighted-sum method in two dimensions, i.e., when $n = 2$. In the two-dimensional space the objective function is a line

$$y = \gamma_1 \cdot f_1(x) + \gamma_2 \cdot f_2(x),$$

where

$$f_2(x) = -\frac{\gamma_1 \cdot f_1(x)}{\gamma_2} + \frac{y}{\gamma_2}.$$

The minimization of $\gamma \cdot f(x)$ in the weight-sum approach can be interpreted as the attempt to find the y value for which, starting from the origin point, the line with slope $-\frac{\gamma_1}{\gamma_2}$ is tangent to the region C.
Obviously, changing the weight parameters leads to possibly different touching points of the line to the feasible region. If the Pareto curve is convex then there is room to calculate such points for different γ vectors (see Fig. 2.3).

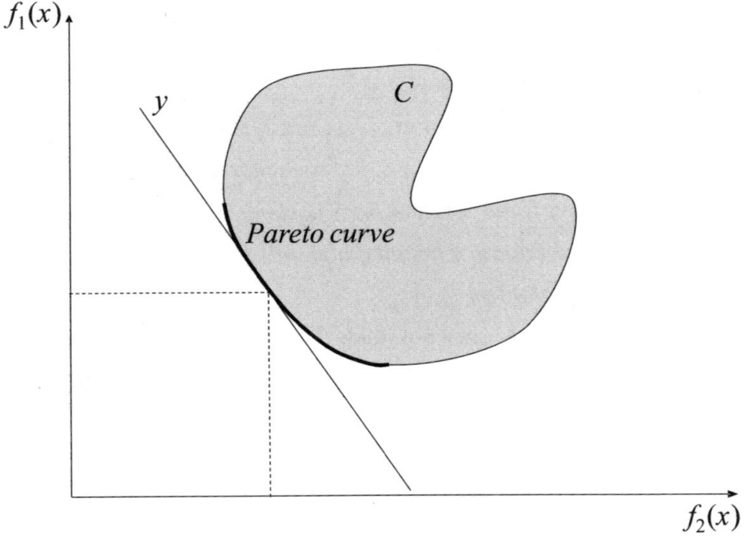

Fig. 2.3 Geometrical representation of the weight-sum approach in the convex Pareto curve case

On the contrary, when the curve is non-convex, there is a set of points that cannot be reached for any combinations of the γ weight vector (see Fig. 2.4).

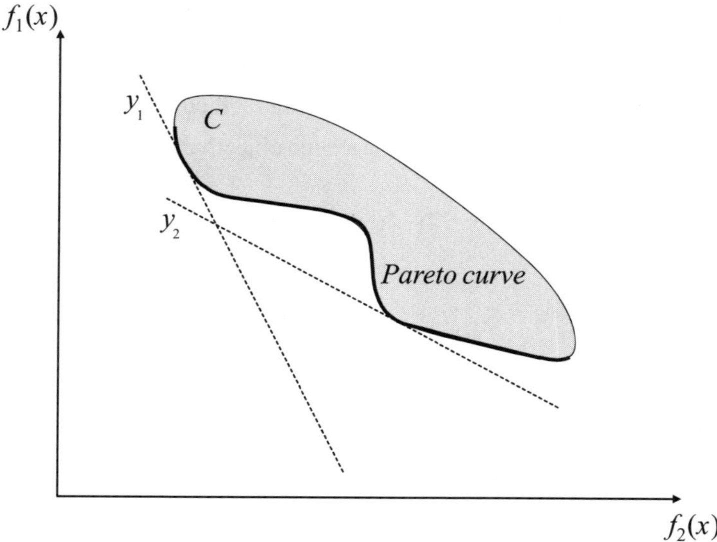

Fig. 2.4 Geometrical representation of the weight-sum approach in the non-convex Pareto curve case

The following result by Geoffrion (1968) states a *necessary and sufficient condition* in the case of convexity as follows:
If the solution set S is convex and the n objectives f_i are convex on S, x^* is a strict Pareto optimum if and only if it exists $\gamma \in R^n$, such that x^* is an optimal solution of problem $P_s(\gamma)$. Similarly: *If the solution set S is convex and the n objectives f_i are convex on S, x^* is a weak Pareto optimum if and only if it exists $\gamma \in R^n$, such that x^* is an optimal solution of problem $P_w(\gamma)$.*

If the convexity hypothesis does not hold, then only the necessary condition remains valid, i.e., the optimal solutions of $P_s(\gamma)$ and $P_w(\gamma)$ are strict and weak Pareto optima, respectively.

2.3.2 ε-constraints Method

Besides the scalarization approach, another solution technique to multi-objective optimization is the *ε-constraints* method proposed by Chankong and Haimes in 1983. Here, the decision maker chooses one objective out of *n* to be minimized; the remaining objectives are constrained to be less than or equal to given target values. In mathematical terms, if we let $f_2(x)$ be the objective function chosen to be minimized, we have the following problem $P(\varepsilon_2)$:

$$\min f_2(x)$$
$$f_i(x) \leq \varepsilon_i, \forall i \in \{1,\ldots,n\} \setminus \{2\}$$
$$x \in S.$$

We note that this formulation of the ε-constraints method can be derived by a more general result by Miettinen, that in 1994 proved that:
If an objective j and a vector $\varepsilon = (\varepsilon_1,\ldots,\varepsilon_{j-1},\varepsilon_{j+1},\ldots,\varepsilon_n) \in R^{n-1}$ exist, such that x^* is an optimal solution to the following problem $P(\varepsilon)$:

$$\min f_j(x)$$
$$f_i(x) \leq \varepsilon_i, \forall i \in \{1,\ldots,n\} \setminus \{j\}$$
$$x \in S,$$

then x^* is a weak Pareto optimum.

In turn, the Miettinen theorem derives from a more general theorem by Yu (1974) stating that:
x^* is a strict Pareto optimum if and only if for each objective j, with $j = 1,\ldots,n$, there exists a vector $\varepsilon = (\varepsilon_1,\ldots,\varepsilon_{j-1},\varepsilon_{j+1},\ldots,\varepsilon_n) \in R^{n-1}$ such that $f(x^*)$ is the unique objective vector corresponding to the optimal solution to problem $P(\varepsilon)$.

Note that the Miettinen theorem is an *easy implementable* version of the result by Yu (1974). Indeed, one of the difficulties of the result by Yu, stems from the uniqueness constraint. The weaker result by Miettinen allows one to use a necessary condition to calculate weak Pareto optima independently from the uniqueness of the optimal solutions. However, if the set S and the objectives are convex this result becomes a necessary and sufficient condition for weak Pareto optima. When, as in problem $P(\varepsilon_2)$, the objective is fixed, on the one hand, we have a more simplified version, and therefore a version that can be more easily implemented in automated decision-support systems; on the other hand, however, we cannot say that in the presence of S convex and f_i convex, $\forall i = 1,\ldots,n$, all the set of weak Pareto optima can be calculated by varying the ε vector.

One advantage of the ε-constraints method is that it is able to achieve efficient points in a non-convex Pareto curve. For instance, assume we have two objective functions where objective function $f_1(x)$ is chosen to be minimized, i.e., the problem is

$$\min f_1(x)$$
$$f_2(x) \leq \varepsilon_2$$
$$x \in S,$$

we can be in the situation depicted in Fig. 2.5 where, when $f_2(x) = \varepsilon_2$, $f_1(x)$ is an efficient point of the non-convex Pareto curve.

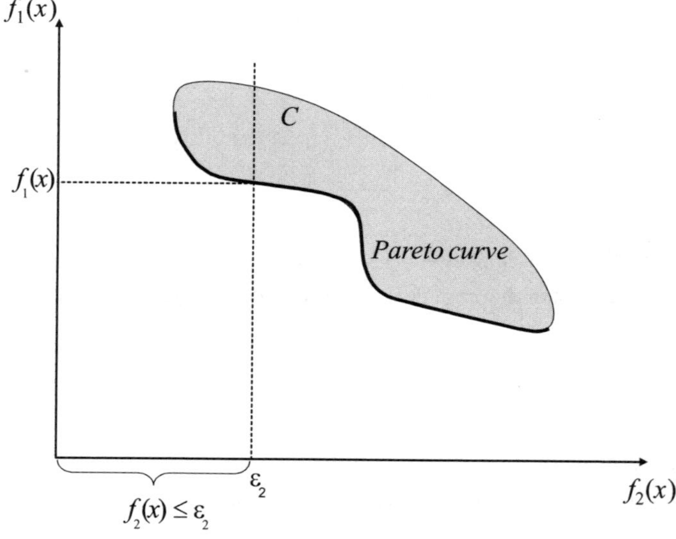

Fig. 2.5 Geometrical representation of the ε-constraints approach in the non-convex Pareto curve case

Therefore, as proposed in Steurer (1986) the decision maker can vary the upper bounds ε_i to obtain weak Pareto optima. Clearly, this is also a drawback of this method, i.e., the decision maker has to choose appropriate upper bounds for the constraints, i.e., the ε_i values. Moreover, the method is not particularly efficient if the number of the objective functions is greater than two.

For these reasons, Erghott and Rusika in 2005, proposed two modifications to improve this method, with particular attention to the computational difficulties that the method generates.

2.3.3 Goal Programming

Goal Programming dates back to Charnes et al. (1955) and Charnes and Cooper (1961). It does not pose the question of maximizing multiple objectives, but rather it attempts to find specific goal values of these objectives. An example can be given by the following program:

$$\begin{aligned}
&f_1(x) \geq v_1\\
&f_2(x) = v_2\\
&f_3(x) \leq v_3\\
&x \in S.
\end{aligned}$$

Clearly we have to distinguish two cases, i.e., if the intersection between the image set C and the *utopian set*, i.e., the image of the admissible solutions for the objectives, is empty or not. In the former case, the problem transforms into one in which we have to find a solution whose value is as close as possible to the utopian set. To do this, additional variables and constraints are introduced. In particular, for each constraint of the type

$$f_1(x) \geq v_1$$

we introduce a variable s_1^- such that the above constraint becomes

$$f_1(x) + s_1^- \geq v_1.$$

For each constraint of the type

$$f_2(x) = v_2$$

we introduce a surplus two variables s_2^+ and s_2^- such that the above constraint becomes

$$f_2(x) + s_2^- - s_2^+ = v_2.$$

For each constraint of the type

$$f_3(x) \leq v_3$$

we introduce a variable s_3^+ such that the above constraint becomes

$$f_3(x) - s_3^+ \leq v_3.$$

Let us denote with s the vector of the additional variables. A solution (x,s) to the above problem is called a strict Pareto-slack optimum if and only if a solution (x',s'), for every $x' \in S$, such that $s_i' \leq s_i$ with at least one strict inequality does not exist.

There are different ways of optimizing the slack/surplus variables. An example is given by the Archimedean goal programming, where the problem becomes that of minimizing a linear combination of the surplus and slack variables each one weighted by a positive coefficient α as follows:

$$\begin{aligned}
&\min \alpha_{s_1^-} s_1^- + \alpha_{s_2^+} s_2^+ + \alpha_{s_2^-} s_2^- + \alpha_{s_3^+} s_3^+ \\
&f_1(x) + s_1^- \geq v_1 \\
&f_2(x) + s_2^- - s_2^+ = v_2 \\
&f_3(x) - s_3^+ \leq v_3 \\
&s_1^- \geq 0 \\
&s_2^+ \geq 0 \\
&s_2^- \geq 0 \\
&s_3^+ \geq 0 \\
&x \in S.
\end{aligned}$$

For the above problem, the Geoffrion theorem says that the resolution of this problem offers strict or weak Pareto-slack optimum.

Besides Archimedean goal programming, other approaches are the lexicographical goal programming, the interactive goal programming, the reference goal programming and the multi-criteria goal programming (see, e.g., T'kindt and Billaut, 2005).

2.3.4 *Multi-level Programming*

Multi-level programming is another approach to multi-objective optimization and aims to find one optimal point in the entire Pareto surface. Multi-level programming orders the n objectives according to a hierarchy. Firstly, the minimizers of the first objective function are found; secondly, the minimizers of the second most important objective are searched for, and so forth until all the objective function have been optimized on successively smaller sets.

Multi-level programming is a useful approach if the hierarchical order among the objectives is meaningful and the user is not interested in the continuous trade-off among the functions. One drawback is that optimization problems that are solved near the end of the hierarchy can be largely constrained and could become infeasible, meaning that the less important objective functions tend to have no influence on the overall optimal solution.

Bi-level programming (see, e.g., Bialas and Karwan, 1984) is the scenario in which $n = 2$ and has received several attention, also for the numerous applications in which it is involved. An example is given by hazmat transportation in which it has been mainly used to model the network design problem considering the government and the carriers points of view: see, e.g., the papers of Kara and Verter (2004), and of Erkut and Gzara (2008) for two applications (see also Chap. 4 of this book).

In a bi-level mathematical program one is concerned with two optimization problems where the feasible region of the first problem, called the upper-level (or leader) problem, is determined by the knowledge of the other optimization problem, called the lower-level (or follower) problem. Problems that naturally can be modelled by means of bi-level programming are those for which variables of the first problem are constrained to be the optimal solution of the lower-level problem.

In general, bi-level optimization is issued to cope with problems with two decision makers in which the optimal decision of one of them (the leader) is constrained by the decision of the second decision maker (the follower). The second-level decision maker optimizes his/her objective function under a feasible region that is defined by the first-level decision maker. The latter, with this setting, is in charge to define all the possible reactions of the second-level decision maker and selects those values for the variable controlled by the follower that produce the best outcome for his/her objective function.

A bi-level program can be formulated as follows:

$$\begin{aligned} &\min f(x_1, x_2) \\ &x_1 \in X_1 \\ &x_2 \in \arg\min g(x_1, x_2) \\ &x_2 \in X_2. \end{aligned}$$

The analyst should pay particular attention when using bi-level optimization (or multi-level optimization in general) in studying the uniqueness of the solutions of the follower problem. Assume, for instance, one has to calculate an optimal solution x_1^* to the leader model. Let x_2^* be an optimal solution of the follower problem associated with x_1^*. If x_2^* is not unique, i.e., $|\arg\min g(x_1^*, x_2)| > 1$, we can have a situation in which the follower decision maker can be free, without violating the leader constraints, to adopt for his problem another optimal solution different from x_2^*, i.e., $\hat{x}_2 \in \arg\min g(x_1^*, x_2)$ with $\hat{x}_2 \neq x_2^*$, possibly inducing a $f(x^*1, \hat{x}_2) > f(x^*1, x_2^*)$ on the leader, forcing the latter to carry out a sensitivity analysis on the values attained by his objective function in correspondence to all the optimal solutions in $\arg\min g(x_1^*, x_2)$.

Bi-level programs are very closely related to the van Stackelberg equilibrium problem (van Stackelberg, 1952), and the mathematical programs with equilibrium constraints (see, e.g., Luo et al. 1996). The most studied instances of bi-level programming problems have been for a long time the linear bi-level programs, and therefore this subclass is the subject of several dedicated surveys, such as that by Wen and Hsu (1991).

Over the years, more complex bi-level programs were studied and even those including discrete variables received some attention, see, e.g., Vicente et al. (1996). Hence, more general surveys appeared, such as those by Vicente and Calamai (1994) and Falk and Liu (1995) on non-linear bi-level programming. The combinatorial nature of bi-level programming has been reviewed in Marcotte and Savard (2005).

Bi-level programs are hard to solve. In particular, linear bi-level programming has been proved to be strongly NP-hard (see, Hansen et al., 1992); Vicente et al. (1996) strengthened this result by showing that finding a certificate of local optimality is also strongly NP-hard.

Existing methods for bi-level programs can be distinguished into two classes. On the one hand, we have convergent algorithms for general bi-level programs with theoretical properties guaranteeing suitable stationary conditions; see, e.g., the implicit function approach by Outrata et al. (1998), the quadratic one-level reformulation by Scholtes and Stohr (1999), and the smoothing approaches by Fukushima and Pang (1999) and Dussault et al. (2004).

With respect to the optimization problems with complementarity constraints, which represent a special way of solving bi-level programs, we can mention the papers of Kocvara and Outrata (2004), Bouza and Still (2007), and Lin and Fukushima

(2003, 2005). The first work presents a new theoretical framework with the implicit programming approach. The second one studies convergence properties of a smoothing method that allows the characterization of local minimizers where all the functions defining the model are twice differentiable. Finally, Lin and Fukushima (2003, 2005) present two relaxation methods.

Exact algorithms have been proposed for special classes of bi-level programs, e.g., see the vertex enumeration methods by Candler and Townsley (1982), Bialas and Karwan (1984), and Tuy et al. (1993) applied when the property of an extremal solution in bi-level linear program holds. Complementary pivoting approaches (see, e.g., Bialas et al., 1980, and Júdice and Faustino, 1992) have been proposed on the single-level optimization problem obtained by replacing the second-level optimization problem by its optimality conditions. Exploiting the complementarity structure of this single-level reformulation, Bard and Moore (1990) and Hansen et al. (1992), have proposed branch-and-bound algorithms that appear to be among the most efficient. Typically, branch-and-bound is used when the lower-level problem is convex and regular, since the latter can be replaced by its Karush–Kuhn–Tucker (KKT) conditions, yielding a single-level reformulation. When one deals with linear bi-level programs, the complementarity conditions are intrinsically combinatorial, and in such cases branch-and-bound is the best approach to solve this problem (see, e.g., Colson et al., 2005). A cutting-plane approach is not frequently used to solve bi-level linear programs. Cutting-plane methods found in the literature are essentially based on Tuy's concavity cuts (Tuy, 1964). White and Anandalingam (1993) use these cuts in a penalty function approach for solving bi-level linear programs. Marcotte et al. (1993) propose a cutting-plane algorithm for solving bi-level linear programs with a guarantee of finite termination. Recently, Audet et al. (2007), exploiting the equivalence of the latter problem with a mixed integer linear programming one, proposed a new branch-and-bound algorithm embedding Gomory cuts for bi-level linear programming.

2.4 Multi-objective Optimization Integer Problems

In the previous section, we gave general results for continuous multi-objective problems. In this section, we focus our attention on what happens if the optimization problem being solved has integrality constraints on the variables. In particular, all the techniques presented can be applied in these situations as well, with some limitations on the capabilities of these methods to construct the Pareto front entirely. Indeed, these methods are, in general, very hard to solve in real applications, or are unable to find all efficient solutions. When integrality constraints arise, one of the main limits of these techniques is in the inability of obtaining some Pareto optima; therefore, we will have *supported* and *unsupported* Pareto optima.

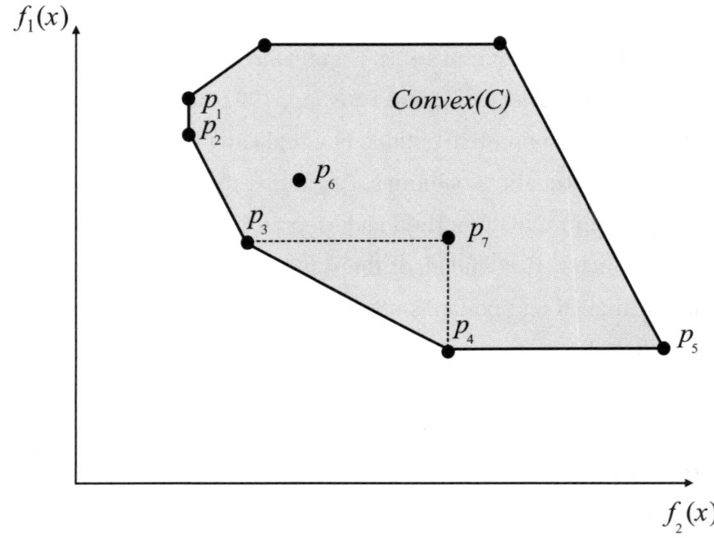

Fig. 2.6 Supported and unsupported Pareto optima

Fig. 2.6 gives an example of these situations: points p_6 and p_7 are unsupported Pareto optima, while p_1 and p_5 are supported weak Pareto optima, and p_2, p_3, and p_4 are supported strict Pareto optima.

Given a multi-objective optimization integer problem (MOIP), the scalarization in a single objective problem with additional variables and/or parameters to find a subset of efficient solutions to the original MOIP, has the same computational complexity issues of a continuous scalarized problem.

In the 2006 paper of Ehrgott "A discussion of scalarization techniques for multiple objective integer programming" the author, besides the scalarization techniques also presented in the previous section (e.g., the weighted-sum method, the ε-constraint method), satisfying the linear requirement imposed by the MOIP formulation (where variables are integers, but constraints and objectives are linear), presented more methods like the Lagrangian relaxation and the elastic-constraints method.

By the author's analysis, it emerges that the attempt to solve the scalarized problem by means of Lagrangian relaxation would not lead to results that go beyond the performance of the weighted-sum technique. It is also shown that the general linear scalarization formulation is NP-hard. Then, the author presents the elastic-constraints method, a new scalarization technique able to overcome the drawback of the previously mentioned techniques related to finding all efficient solutions, combining the advantages of the weighted-sum and the ε-constraint methods. Furthermore, it is shown that a proper application of this method can also give reasonable computing times in practical applications; indeed, the results obtained by the author on the elastic-constraints method are applied to an airline-crew scheduling problem, whose size oscillates from 500 to 2000 constraints, showing the effectiveness of the proposed technique.

2.4.1 Multi-objective Shortest Paths

Given a directed graph $G = (V,A)$, an origin $s \in V$ and a destination $t \in V$, the *shortest-path problem* (SPP) aims to find the minimum distance path in G from o to d. This problem has been studied for more than 50 years, and several polynomial algorithms have been produced (see, for instance, Cormen et al., 2001).

From the freight distribution point of view the term *shortest* may have quite different meanings from faster, to quickest, to safest, and so on, focusing the attention on what the labels of the arc set A represent to the decision maker. For this reason, in some cases we will find it simpler to define for each arc more labels so as to represent the different arc features (e.g., length, travel time, estimated risk).

The problem to find multi-objective shortest paths (MOSPP) is known to be *NP*-hard (see, e.g., Serafini, 1986), and the algorithms proposed in the literature faced the difficulty to manage the large number of non-dominated paths that results in a considerable computational time, even in the case of small instances. Note that the number of non-dominated paths may increase exponentially with the number of nodes in the graph (Hansen, 1979).

In the multi-objective scenario, each arc (i,j) in the graph has a vector of costs $c_{ij} \in R^n$ with $c_{ij} = (c^1_{ij}, \ldots, c^n_{ij})$ components, where n is the number of criteria.

A path P_{si} from the origin s to node i is a sequence of nodes (and arcs) $P_{si} = (s \equiv n_1, \ldots, n_h \equiv i)$ of length $h \geq 2$, where each arc $(n_l, n_{l+1}) \in A$ for $l = 1, \ldots, h-1$. Such a path is evaluated by means of a performance vector $c(P_{si}) = (c^1(P_{si}), \ldots, c^n(P_{si}))$ where

$$c^l(P_{si}) = \sum_{(p,q) \in P_{si}} c^l_{pq}$$

for $l = 1, \ldots, n$.

Let

$$x_{ij} = \begin{cases} 1 \text{ if } (i,j) \text{ is in a path;} \\ 0 \text{ otherwise.} \end{cases}$$

The problem to find the multi-objective shortest paths from the origin s to destination t could be stated as follows:

$$\min f^l(x) = \sum_{(i,j) \in A} c^l_{ij} x_{ij} \forall l \in \{1, \ldots, n\}$$

$$\sum_{\{j:(i,j) \in A\}} x_{ij} - \sum_{\{j:(j,i) \in A\}} x_{ji} = \begin{cases} 1 & i = s \\ 0 & \forall i \in V \backslash \{s,t\} \\ -1 & i = t \end{cases}$$

$$x_{ij} \geq 0, \ \forall (i,j) \in A.$$

Let the solution space of the above problem be denoted with S_{st}. Let P^1_{st} and P^2_{st} be two feasible paths and $c(P^1_{st})$ and $c(P^2_{st})$ be the relative performance vectors.

Given a path $P^1_{st} \in S_{s,t}$, the vector $c(P^1_{st})$ is *non-dominated* iff there does not exist another vector $c(P^2_{st})$, with $P^2_{st} \in S_{s,t}$, such that $c^l(P^2_{st}) \leq c^l(P^1_{st})$, $l = 1, \ldots, n$ and $c^l(P^2_{st}) \neq c^l(P^1_{st})$ for some l; otherwise, $c(P^2_{st})$ *dominates* $c(P^1_{st})$.

Given a path $P^1_{st} \in S_{s,t}$, the vector $c(P^1_{st})$ is *weakly non-dominated* if and only if there does not exist another vector $c(P^2_{st})$, with $P^2_{st} \in S_{s,t}$, such that $c^l(P^2_{st}) < c^l(P^1_{st})$, $l = 1, \ldots, n$; otherwise, $c(P^2_{st})$ *strictly dominates* $c(P^1_{st})$.

Correspondingly, efficient solutions are defined as follows: a path P^1_{st} is strict efficient (or strict Pareto optimal) if only if $P^1_{st} \in S_{st}$ and there does not exist another path $P^2_{st} \in S_{s,t}$ such that $c^l(P^2_{st})$ dominates $c^l(P^1_{st})$.

A path P^1_{st} is weakly efficient if and only if $P^1_{st} \in S_{st}$ and there does not exist another path $P^2_{st} \in S_{s,t}$ such that $c^l(P^2_{st})$ strictly dominates $c^l(P^1_{st})$.

Hansen's (1980) work was one of the first studies on the multi-objective shortest-path problem; he analyzed some bi-objective problems and demonstrated that, in the worst case, the non-dominated paths number grows exponentially with the network size. So it is clear that, in the worst case, the generation of the efficient frontier may require a huge computational effort. A way to find a subset of non-dominated solutions can be to use a linear combination of the objectives in question, changing the used weights. On this subject, Henig (1986) proposed a review of exact methods for finding all non-dominated paths, and presented an approximated method whose average computational complexity is polynomial in the size of the network nodes.

Even if there are some dynamic programming approaches (e.g., in Henig, 1986, and Kostreva and Wiecek, 1993), the study on algorithms for the multi-objective shortest-path problem was mainly devoted to two classes of methods: one based on label-setting algorithms and the other based on label-correcting algorithms. In the former class, most of the work done was devoted to problems where all the objective functions are of the sum type, as in Hansen (1979), and Martins (1984).

Martins (1999) proposed a multi-objective algorithm based on vertex labelling techniques that generalizes the Bellman optimality principle. In his paper he assumed the attributes associated on arcs to be non-negative, deterministic and additive along the route. Martins' algorithm uses a multiple labelling approach. Each node $i \in V$ is associated with several labels, and the lth label contains the n objective values and two pointers. The label can be represented as

$$[c^1(P_{si}),\ldots,c^r(P_{si}),j,l_1]_l,$$

where $c^h(P_{si})$ is the length of the path P_{si} from origin s to node i, for $h = 1,\ldots,n$, $j \neq i$ is some node of G, and l_1 indicates a certain label of node j for which

$$c_{[h,l]}(P_{si}) = c_{[h,l_1]}(P_{sj}) + c^h_{ij},$$

where $c^{P_{si}}_{[h,l]}$ is the hth component of the lth label of node i.

At each iteration, there are two kinds of labels, permanent ones and temporary ones. The algorithm selects a temporary label in a node i, converts it to permanent, and updates all the labels of the successors j of i, for each $(i,j) \in A$; then, it deletes all the labels that represent a dominated path P_{sj}. The algorithm stops when it runs out of temporary labels, and each permanent label represents a unique efficient path.

The node-selection step is made considering, among all the labels in each node, the lexicographically smallest one. We recall that for some node i, a label

$$[c^1(P_{si}),\ldots,c^n(P_{si}),-,-]_\xi$$

is said to be lexicographically smaller than a label

$$[c^1(P'_{si}),\ldots,c^n(P'_{si}),-,-]_\delta$$

if $c_{[1,\xi]}(P_{si}) = c_{[1,\delta]}(P'_{si}),\ldots,c_{[k-1,\xi]}(P_{si}) = c_{[k-1,\delta]}(P'_{si})$, $c_{[k,\xi]}(P_{si}) < c_{[k,\delta]}(P'_{si})$ holds for some $k \in \{1,\ldots,n\}$.

In Table 2.1, Martins' algorithm is sketched.

Table 2.1 Martins' algorithm

Step 1	Assign the temporary label $[(0,0,\ldots,0),-,-]_1$ to node s.
Step 2	If the set of temporary labels is empty go to step 5. Otherwise, among all the temporary labels determine the lexicographically smallest one. Let the lth label be associated with node i. Set this label as a permanent one.
Step 3	While some node $j \in V$ exists, such that $(i,j) \in A$, execute
	Step 3.1 $c^k(P_{sj}) = c_{[k,l]}(P_{si}) + c^k_{ij}$ for every $k = 1,\ldots,n$ and let $[(c^1(P_{sj}),\ldots,c^n(P_{sj})),i,l]_\xi$ be a new temporary label of node j.
	Step 3.2 Among all the temporary labels of node j, delete all labels representing a dominated path from s to j.
Step 4	Return to step 2.
Step 5	Find the non-dominated paths from s to t. For that, the two pointers of each label must be used.
Step 6	Stop.

Recently, Gandibleux et al. in 2006, proposed a problem with a max-min objective and two min-sum objectives that was solved using a revisited version of Martins' algorithm. Referring to the class of label-setting algorithms, where node labels are marked as permanent only at the last iteration of the run, we refer the reader to, e.g., Brumbaugh-Smith and Shier (1989), Mote et al. (1991), and Skriver and Andersen (2000).

For the bi-objective case, there are algorithms for the path-ranking problem, as in Climaco and Martins (1982) and Martins and Climaco (1981), and for the max-ordering problem, as in Ehrgott and Skriver (2003). In Paixão et al. (2003) ranking is obtained for the general multi-objective shortest-path problem using a utility function based on the norm value associated with each path.

As described in Tarapata (2007), one of the most popular methods of solving multi-objective shortest-path problems is the construction of $(1+\varepsilon)$-Pareto curves (see, e.g., Papadimitriou and Yannakakis, 2000, and Vassilvitskii and Yannakakis, 2004). Informally, a $(1+\varepsilon)$-Pareto curve P_ε is a subset of feasible solutions such that for any Pareto-optimal solution there exists a solution in P_ε that is no more than $(1-\varepsilon)$ away in the objectives (Tarapata, 2007). Papadimitriou and Yannakakis (2000) show that for any multi-objective optimization problem there exists a $(1+\varepsilon)$-Pareto curve P_ε of (polynomial) size. Extensions to this method to produce a constant approximation to the smallest possible $(1+\varepsilon)$-Pareto curve for the cases of 2 and 3 objectives are presented in Vassilvitskii and Yannakakis (2004), while for a number of objectives greater than three, inapproximability results are shown for such a constant approximation.

For the case of the multi-objective shortest-path problem (and some other problems with linear objectives), Papadimitriou and Yannakakis (2000) show how a gap routine can be constructed (based on a pseudopolynomial algorithm for computing exact paths) and, consequently, provide a fully polynomial time approximation scheme for this problem. Note that fully polynomial time approximation schemes for the multi-objective shortest-path problem were already known in the case of two objectives (Hansen, 1979), as well as in the case of multiple objectives in directed acyclic graphs (Warburton, 1987). In particular, the bi-objective case was extensively studied (see, e.g., Ehrgott and Gandibleux, 2002), while for a number of objectives greater than two, very little has been achieved. The results in Warburton (1987), Papadimitriou and Yannakakis (2000) and Tsaggouris and Zaroliagis (2005) are the best fully polynomial time approximation schemes known.

Warburton (1987) has proposed polynomial methods for generating all "quasi" Pareto-optimal paths. Another approach can be to treat each objective as a constraint: Sancho (1988) has considered the problem with three objectives, two of which have been transformed into constraints. Each of the generated solutions cannot be Pareto optimal with respect to all three considered objectives.

An approach not using deterministic attributes was proposed by List et al. (1991). In particular, they developed a method that finds, given an origin and a destination, the non-dominated paths replacing to each stochastic attribute (supposed normally distributed) its mean and variance; then, the non-dominated paths number is reduced by comparing all pairs of paths, using the "stochastic dominance" criterion.

2.4.2 *Multi-objective Travelling Salesman Problem*

The *travelling salesman problem* (TSP) is different from the shortest-path problem, since here the node set V of the graph $G = (V,A)$ must be visited once and exactly once. If arcs are labelled with arc lengths we seek the tour of minimum total length. We remark again, as for the SPP, that labels on the arcs can have different meanings. For a recent survey on TSP and its variations, the reader is referred, e.g., to the book "The Travelling Salesman Problem and its Variations" by Gutin and Punnel (2006).

Manthey and Ram proposed in 2006 an excellent state-of-the-art analysis on the multi-objective travelling salesman problem. Also, Ehrgott in 2000 developed a relevant survey on the travelling salesman problem with multiple criteria in his paper on "Approximation algorithms for multi-criteria combinatorial optimization problems". Beyond the two mentioned papers, important results on the multi-criteria TSP have been achieved by Angel et al. (2004, 2005). In this section, following the latter papers, we recall the main results in this area.

Ehrgott (2000) analyzed a generalization of Christofides' algorithm for the TSP with triangle inequalities and the symmetry property, i.e., Δ-STSP. Instead of considering Pareto curves, he measured the quality of a solution as a norm of the objective function vector, i.e., scalarizing the objectives in a single one. The approximation ratio achieved is between $\frac{3}{2}$ and 2, depending on the norm used to combine the different criteria.

Angel et al. (2004) considered the two-criteria symmetric TSP with weights one and two. They presented a $\frac{3}{2}$-approximation algorithm for this problem by using a local search heuristic.

Successively, Angel et al. (2005) generalized these results to the k-criteria symmetric TSP with weights one and two by presenting a $2 - \frac{2}{k+1}$-approximation, with $k \geq 3$.

More recently, in Manthey and Ram (2006), the authors presented a new approximation results for the multi-objective symmetric TSP of $(1 - \gamma - \varepsilon)$-approximation with $\gamma \in [\frac{1}{2}, 1]$.

2.4.3 Other Work in Multi-objective Combinatorial Optimization Problems

In the 2006 paper of Ehrgott et al. “A level set method for multi-objective combinatorial optimization: application to the quadratic assignment problem”, the authors studied the assignment problem in the multi-objective case.

The basic version of the well-known *assignment problem* (AP) can be defined on a bipartite graph. One may be interested in finding the maximum number of edges that do not have a common vertex: when AP is used, e.g., to model the problem of assigning loads to trucks or people (drivers) to vehicles this corresponds to finding the maximum number of loads assigned to trucks (where the edge between a load and a truck means compatibility). In other cases edges can be labelled, for instance, with an estimation of the payoff obtained by a certain assignment, and then the goal is to search for the set of edges with no vertex in common that maximizes the total estimated payoff. Similarly, if labels on edges represent a certain cost one may also seek for an assignment that minimizes the total cost. The reader can find algorithms and variations on this problem, e.g., in “Network Flows: Theory, Algorithms, and Applications”, by Ahuja et al. (1993).

Studying a quadratic version of the assignment problem (QAP) with multiple objectives, Erghott et al. (2006) pose a twofold goal: on the one hand, developing a procedure for the determination of Pareto-optimal solutions in a multi-criteria combinatorial optimization problem; on the other hand, demonstrating the effectiveness of the previous method on QAP. The method proposed to accomplish the former goal is based on a result due to Ehrgott et al. (1997). The routine, based on the notion of level sets and level curves, works with an arbitrary number of objectives, and uses an algorithm that solves the problem of finding a K best solution in a (single-objective) combinatorial optimization problem. Regarding the second goal, i.e., solving the multi-objective quadratic assignment problem (MQAP), the proposed approach represents the first method able to handle the MQAP. Furthermore, the authors present two algorithms for ranking QAP solutions and report an experimental comparison between them on randomly generated examples with a number of objective functions ranging from 2 to 6.

2.5 Multi-objective Combinatorial Optimization by Metaheuristics

In this section, we present the state-of-the-art on metaheuristics to solve multi-objective combinatorial optimization problems. A comprehensive study on multi-objective optimization by metaheuristics can be found in the 2001 book "Multiple Objective Metaheuristic Algorithms for Combinatorial Optimization" of Andrzej Jaszkiewicz.

The first multiple objective metaheuristic proposed in the literature was due to Schaffer in 1985 (vector evaluated genetic algorithm, VEGA). It is an adaptation of a single-objective genetic algorithm with a modified selection mechanism. It was observed by Coello (1999) that the solutions generated by VEGA are in general not uniformly spread in the Pareto front; indeed, they tend to poorly represent the middle regions of the non-dominated points set. Fonseca and Fleming (1995) and Tamaki et al. (1996) proposed a modification of this method in order to circumvent such a shortcoming. More recently, evolutionary algorithms have also been tailored to solve optimization problems with multiple conflicting objectives, by approximating the Pareto set in such problems. A complete tutorial on evolutionary multi-objective optimization can be found in the papers of Zitzler et al. (2004) and Deb (2001, 2005).

A genetic/evolutionary algorithm operates on a set of candidate solutions that is subsequently modified by two basic operators: *selection* and *variation*. Selection is in charge to model the reproduction mechanism among living beings, while variation mimics the natural capability of creating new living beings by means of recombination and mutation. Two goals have to be taken into account when designing a multi-objective genetic/evolutionary algorithm: guiding the search towards the Pareto set and keeping a diverse set of non-dominated solutions. The first goal is mainly related to mating selection, in particular to the problem of assigning scalar fitness values in the presence of multiple objectives. The second goal concerns selection in general because we want to avoid the situation that the population contains mostly identical solutions (with respect to the objective space and the decision space). Finally, there is a third issue that addresses both of the above goals, i.e., elitism that prevents non-dominated solutions from being lost.

In contrast to single-objective optimization, where objective function and fitness function are directly and easily related, in multi-objective optimization fitness assignment and selection have to take into account all the different objectives. Among the different fitness assignment strategies, the most commonly used are those based on aggregation, single-objective, and Pareto dominance.

The first approach, which mimics the weighted-sum method, aggregates the objectives into a single parameterized objective function. The parameters of this function are systematically varied during the optimization run in order to find a set of non-dominated solutions instead of a single trade-off solution. In order to spread the search over all regions of the non-dominated set various objective weights should be used. Hajela and Lin (1992) proposed to encode the weights in the solutions description. Thus, the weights evolve in results of recombination and mutation operators. Fitness sharing is used in order to achieve diversification of weights. The weights are used in linear scalarizing functions. Murata et al. (1996) proposed to draw at random a weight vector for use in each iteration composed of a single recombination.

Single-objective-based methods switch between the objectives during the selection phase. Each time an individual is chosen for reproduction, potentially a different objective will decide which member of the population will be copied into the mating pool.

The Pareto dominance (or ranking) strategy is based on exploiting the partial order on the population. Some approaches use the dominance rank, i.e., the number of individuals by which an individual is dominated, to determine the fitness values; others make use of the dominance depth, where the population is divided into several fronts and the depth reflects to which front an individual belongs. Alternatively, the dominance count, i.e., the number of individuals dominated by a certain individual, can also be taken into account. Independent of the technique used, the fitness is related to the whole population, in contrast to aggregation-based methods that calculate an individual raw fitness value independently of other individuals. Pareto ranking alone, however, does not guarantee that the population will spread uniformly over the non-dominated points set. It is known that in the case of Pareto ranking-based selection schemes finite populations converge to a single optimum, a phenomenon known as *genetic drift* (Goldberg and Segrest, 1987), implying a convergence to small regions of the Pareto-optimal set. Some authors worked in the direction of generating implementations able to reduce this effect (see, e.g., Fonseca and Fleming, 1995, and Srinivas and Deb, 1994).

Also simulated annealing has been used to cope with multiple objective combinatorial optimization problems. Serafini, in 1992, was the first author to propose such an approach. The method works in a similar way with respect to the standard (single-objective) simulated annealing implementation, and produces a set of (potentially Pareto-optimal) solutions not dominated by any other solution generated by the algorithm so far. In single-objective simulated annealing a new solution whose value is not worse than the current solution is accepted with probability equal to one, while it is accepted with a probability lower than one if it has a worse value than the current solution. In the multi-objective scenario, three situations can occur when comparing the new solution s' with the current solution s: s' dominates s, s' dominates s, s and s' are mutually non-dominated. In the first case, s' is accepted with probability equal to one; in the second case the acceptance probability is lower than one; in the third case Serafini proposed different rules corresponding to some weighted scalarizing functions, showing that these rules guarantee that the algorithm achieves one of the Pareto-optimal solutions if the temperature is decreased sufficiently slowly; while they do not guarantee appropriate dispersion over the whole set of non-dominated points. To achieve dispersion the author proposed to modify the weights randomly over time. Also, Ulungu et al. (1999) proposed multi-objective acceptance rules by using predefined weight vectors, where each of the vectors is associated with an independent annealing process. As in the case of the algorithm of Serafini (1992), the algorithm of Ulungu et al. (1999) produces the set of potentially Pareto-optimal solutions containing all the solutions not dominated by any other solution generated by any of the annealing processes. The reader can also refer to the paper of Suppapitnarm and Parks (1999) to see another implementation of simulated annealing for multi-objective problems.

Gandibleux et al. (1997) proposed a multi-objective version of tabu search. The method is based on a periodical modification of the weights to be assigned to the scalarized objective function, and makes use of two tabu lists, one that prevents already visited solutions from being reconsidered, and the other related to weight vectors. Also, Hansen in 1998 developed a multi-objective tabu search based on assigning weights to the scalarized objectives; the method borrows some ideas from Pareto-simulated annealing since it uses the same solutions dispersion mechanism achieved by a proper modification of the weight vectors.

Chapter 3
Maritime Freight Logistics

Abstract In this chapter, we introduce some aspects of freight distribution problems inherent to a maritime terminal, which represents the origin of the road and the rail shipments. This analysis also has the objective of introducing how simulation tools can be used to set capacity and service level.

3.1 Capacity and Service Level in a Maritime Terminal

Planning and management of a freight distribution network requires that several functional parameters have to be known for each component of the network.

Even if in mathematical optimization models we often assume that data and parameters are given and are quite reliable, we are aware that this is not always the case and that some components of the network in the absence of some parameters have to be analyzed separately also because internal operational rules may change the performance of the system. As a practical example, in this chapter, we show how a maritime container terminal can be analyzed by means of a simulation model and, in the conclusion of this chapter, we will explain how this simulation approach can be integrated with the optimization one.

When the component to be studied is a logistic platform or a maritime terminal, some of these parameters may be derived from the historical data, and should be seen just as estimations to be interpreted according to the conditions in which the system is operating. Indeed, the more complex the platform/terminal, i.e., made of several interconnected components, the more difficult is the estimation of its capacity as a node of a freight distribution network and would require a thorough analysis on its own.

Shipping, with particular reference to containerized shipping, is a competitive service industry. A container terminal (or terminal for short) in a port is the place where container vessels dock in berths and unload inbound (import) containers (empty or filled with cargo) and load outbound (export) containers. The terminals have storage yards for the temporary storage of these containers.

In the book "Container Terminals and Automated Transport Systems" (2005), the editors Günther and Kim pointed out the existence of three different types of problems arising in a container terminal that draw the attention for a quantitative analysis:

- *Design problems* that account for the determination of, e.g., the handling equipment in the yard, the number of berths, quay cranes, yard cranes, storage areas, and human workforce. Here, one is interested in estimating performance measures to evaluate the effectiveness of the terminal configuration. Simulation is one of the main tools to cope with such problems.
- *Operational planning problems*; because of the scarce resource availability in the terminal (e.g., limited number of berths, quay cranes, yard cranes, yard space, and human workforce), scheduling the handling operations in container terminals has to carried out in order to maximize the efficiency of the operations, preventing possibly costly conflicts among jobs.
- *Real-time control problems*; even if resource allocation, for resources like berths, quay cranes, and storage areas, is carried out in a planning phase preceding the usage of the resources themselves, it can happen that adjustments have to be executed in real time, especially in the case of short-term planning.

In this chapter, we will discuss the first two problems mentioned above, pointing out, in the first part, how to measure the performance of the terminal by means of simulation, and, in the second part, how multi-objective optimization comes into play in the operational planning phase.

The competitiveness of a container terminal is strictly related to the service level offered to shipping lines, so the ultimate goal in terminal management is to minimize the ship turnaround time (also called the time in port of the ship). In more detail, the port operator has service contracts with a vessel that promises a maximum waiting time for that vessel, the maximum time between when the vessel arrives at the port and when the vessel is berthed at the terminal. A vessel is said to be berthed-on-arrival if the actual berthing time is within 2h from the arrival. We remark that minimizing the waiting time is an issue for both the vessel and the port operator.

To run terminal operations it is not possible to make use of "day-to-day" planning because vessels and freight dimensions need very long processing times at their berths. Indeed, even before serving a ship for unloading and loading operations, docking a big vessel requires the use of further resources such as a roadstead, a big sheltered creek where arriving vessels can anchor while waiting for a berth, and tugs to pull vessels to their berths.

Nevertheless, it is important to note that the vessel arrival at the port is not treated as a random event. Firstly, there is the "long-term schedule", a monthly plan that includes the following information: line name, ship name, voyage number, ports of shipment (loading/unloading) and scheduled date of arrival. The long-term schedule is followed by the "short-term schedule" that is a weekly plan that provides the extended arrival time (EAT) – the scheduled arrival time and the scheduled containers handling (in/out). Finally, there is the daily planning that is an integration of weekly planning and the previously mentioned information. Therefore, the shipping lines must communicate to the terminal within a twenty-four hour period their final EAT, which includes the line name, ship name, voyage number, ports of shipment, real containers handling (in/out), connection ships' name, voyage number and number of containers (in/out). The terminal needs this information to establish the time and the berth of the vessel taking into account possible variations. Vessels communicate their EAT to the terminal 7–8 days in advance, possibly updating it until 24 h before their arrival at the port. A possible overdue could obviously depend on both the state of the atmosphere (e.g., rough sea, high winds) and on the efficiency of the port of origin.

It is obvious that port dimensions and features determine the amount of container trade in transit at that port. Hence, the simulation of the operations in a port is based on the assumption that quay cranes are bottleneck resources: the credibility of these resources is to be considered fundamental because one of the most important key performance indicators to estimate the productivity of a port is the quay-crane (QC) rate. In the next paragraph, we will use simulation to calculate the amount of freight (in/out) and the resource utilization to draft a procedure to calculate the QC rate and estimate the productivity of a port.

3.1.1 The Simulation Setting

Recalling that it is possible to know the arrival times of the vessels by means of the EAT information, in the following we show a discrete time events simulation model to study the port efficiency problem. The simulation has been implemented by means of the Rockwell Software Arena 5.0 (www.arenasimulation.com). The system we study is a little "transshipment" port of which the service to arriving cargo vessels is simulated by using quay cranes as resources operating on vessel charge and discharge. The problem is simplified by neglecting the resources roadstead and tug: the reason for this choice is that our objective is to estimate the port productivity, so we do not consider resources that do not affect the calculation of the QC rate, in order not to increase the complexity of the model structure. The simulation replication length is one week, i.e., 168 h. An hour is the simulation time unit.

All containers considered in the simulation have the same size, i.e., they are all 1 TEU (twenty-foot equivalent unit) containers (see Fig. 3.1); in this way each vessel arriving or leaving the port has an average cargo, equivalent to the capacity of a single ship. We note that containers can also have a 2 TEU size, i.e., forty-foot size.

Fig. 3.1 TEU containers

The vessels considered in the simulation are cargo ships of lo-lo (lift on–lift off) type, with vertical charge and discharge handled by means of quay cranes.

Liner ships generally operate on closed courses and often no origin–destination route can be defined because the ships must be unloaded and then reloaded each time they call at a port, as they can never voyage empty.

Cargo ships have holds opportunely designed to aid container storage and charge–discharge operations. Shipping with cargo ships and high-skilled container terminals within the ports have allowed a big reduction in handling time, so a further inter-modality level has grown up: ship-to-ship inter-modality. Consequently cargo ships have been divided into two classes, feeder ships and big transoceanic ships: the first ones operate on short routes among transhipment ports (hub) to collect charges destined for other small local ports for inland distribution, the second ones only call at transshipment ports.

Big transoceanic vessels are usually divided into two categories: panamax (compatible to cross Panama Canal) and postpanamax (not compatible to cross Panama Canal). Even if transshipment field evolution drives shipping companies to a "naval gigantism" (more than 6000 TEU capacity vessels are already operational, and 10000 TEU capacity vessels are being planned), in this simulation problem we will consider the following two kinds of ships:

- feeder ships,
- panamax ships,

whose sizes are indicated in Table 3.1.

Table 3.1 Sizes of panamax and feeder ships

Features	Unit	Feeder	Panamax
Length	Mt	87	210–290
Width	Mt	15–22	22–32
Draught	Mt	12	12

Vessel capacities are indicated in Table 3.2. Notice that panamax ships arriving at the port will discharge the whole amount of freight (2000 TEU) and then they will be completely re-loaded; feeder ships instead arrive empty at the port and will, therefore, only be completely loaded (500 TEU). Resources used to serve the ships are the quays (berthing points) and the quay cranes.

Table 3.2 Transportable TEU per vessel

Vessel	Feeder	Panamax
Capacity	500 TEU	2000 TEU

Quay is used to simulate the berth assignment to an arriving ship, that holds the berth and quay cranes at the same berth: docking requires about one hour and unloading and loading operations at the berth can start immediately. Table 3.3 refers to quay-crane average operations times used for the simulation: it shows the time required for loading and unloading operations according to the two types of vessels considered, i.e., panamax and feeder.

Table 3.3 Quay-crane average operations time (s) per container (1 TEU). *Average value per operation per vessel. **Average value per vessel

Operation	Load		Unload	
Vessels	Panamax	Feeder	Pamamax	Feeder
Coupler* (s)	9–13	9–13	9–13	–
Lifting** (s)	18–24	12–16	18–24	–
Translation** (s)	17–21	11–14	17–21	–
Going down** (sc)	18–25	12–16	18–25	–
Release* (s)	8–12	8–12	8–12	–
Return** (s)	45–53	31–38	45–53	–
Average (s)	116–148	83–109	116–148	–

Quay-crane operation time is fundamental to defining the processing times of the ships: it is defined by a triangular distribution, with lower limit, mode and upper limit values equal to 116 s, 132 s, 148 s, respectively, per quay crane per TEU per panamax, and equal to 83 s, 96 s, 109 s, respectively, per quay crane per TEU per feeder.

Therefore, it is possible to get to processing times used in the simulation, that are triangular distributions defined in the following way:

- lower limit, mode and upper limit equal to 64 h, 73 h, 82 h, respectively, per crane per operation (load or unload) per panamax (equivalent to 2000 TEU);
- lower limit, mode and upper limit equal to 11 h, 13 h, 15 h, respectively, per crane per operation (load) per feeder (equivalent to 500 TEU).

The quay crane is the resource that will allow the berth sizing (in terms of number of quay cranes), in order to minimize the ships turnaround time and the waiting time to dock. To estimate the port productivity, we will calculate, exploiting the simulation results, the average QC rate, which is the quay-crane throughput measure during a period, defined as follows:

$$QC_{\text{rate}} = \frac{\text{\# of containers unloaded, loaded}}{\text{total \# of QC hours of all QCs that worked}} [\text{TEU/h}]$$

As terminals charge shipping companies for every container loaded into or unloaded from a vessel, the average QC rate is a measure to be maximized, in relation to resources used. Note that closely related to the QC rate is the ship turnaround time, and these two measures are used internally by a terminal to calibrate its own performance. In the same terminal, with the same equipment, as the QC rate increases, the ship turnaround time decreases, and viceversa. However, if the terminal acquires new quay cranes and puts them into use, the ship turnaround time may decrease while the QC rate remains unchanged.

3.1.2 The Simulation Model

In the simulation model we will consider two kinds of ships, i.e., feeder and panamax, and four berths, i.e., three for panamax ships and one for feeder ships with four quay cranes at each berth serving panamax ships and one quay crane at the feeder berth.

Activities at the transshipment port are simulated generating panamax ship arrivals, the occupation of the dedicated berths where the four quay cranes are operating to serve loading and unloading operations to the berthed ships, and considering the whole freight as unitary (2000 TEU per vessel is the amount of cargo to unload from the ship and immediately load into the same ship). Finally, as soon as the loading operation is completed, ships will leave the port leaving the berth free for another ship. Moreover, to complete the transshipment context, feeder-ship arrivals are generated; they will arrive empty at the port and will be processed by one quay crane for loading operations with a 500 TEU cargo at the single feeder berth.

Going into detail, the data problems are listed as follows:

- replication length: one week, equivalent to 168 h;

- arrivals: two panamax ships every 24 h, one feeder ship every 12 h;
- service:
  - four quay cranes operating in three berths dedicated to panamax ships (berth A, berth B, berth C), with processing times that follow triangular distributions, with parameters 16 hours, 18.25 h and 20.5 h, per operation (loading/unloading) per ship. Note that four quay cranes at one berth will effect both discharge operation and charge operation on the same berthed ship: thus, resources will not be available to other ships until the charge operation is completed;
  - one quay crane operating in the only berth (berth D) dedicated to feeder ships, with a processing time that follows a triangular distribution, with parameters 11 h, 13 h, and 15 h, per load operation per ship because a feeder arrives empty at the port;
- queue: ships are serviced on a first-come-first-served (FCFS) basis, moreover, panamax ships are sent to their assigned berths on a smaller queue basis at each berth.

Another type of problem to solve in the simulation is to schedule the 2000 TEU freight composition and send it to the berthed panamax ship that is already unloaded. This could be solved by organizing the freight according to ship arrivals: as the arrival times are known, it is possible to prepare and to compose the amount of freight in a specular way with the arrival of the ship that has to embark these containers. Thus, after the unloading operation, when the unloaded freight is sent to the storage yard, it is possible to simulate a short wait in the yard, equivalent to 0.5 h, that is the time necessary to compose the freight to load and to rearrange the quay at the berth, which can be ready exactly on time for the beginning of a loading operation. This device is used in the simulation to avoid long queues and reduce waiting time for outbound containers. Possible delays at the quay before loading operations may be caused by processing time distributions. Another thing to be simulated is the time the ship needs to dock and occupy the berth when it arrives at the port: a panamax ship needs 1 h to berth while a feeder needs 0.5 h.

Running the simulation, we got the following results:

- Entities - number in: 16 panamax ships and 15 feeder ships. Number of entities generated during the simulation replication length (168 h).

- Entities - number out: 10 panamax ships and 12 feeder ships. Number of entities that have been disposed of, i.e., that have left the system. Note that ships leave the port with their freights (2000 TEU per panamax ship and 500 TEU per feeder ship).
- Entities in the system: 6 panamax ships and 3 feeder ships. Number of entities that have remained in the system when the simulation is over. There are 3 panamax ships in the queue waiting for a berth, one waiting at berth A, one at berth B and one at berth C. There are 1 panamax ship being charged at berth A, 2 panamax ships being discharged, one at berth B, and one at berth C. While there is 1 feeder ship in the process of being charged at berth D and 2 feeder ships in queue waiting for the same berth D.
- Stored cargos: 11 2000 TEU container freights unloaded from panamax ships and sent to the storage yard, 4 arriving from berth A, 4 arriving from berth B and 3 arriving from berth C. The total amount of stored cargos is equivalent to 22,000 TEU.
- Average total time spent in the system by the entities, from their arrival at the port until their exit; it includes the processing times for loading and unloading operations, the processing times to dock and the waiting times to have access to berths:
  - Panamax: 57.9 h (maximum value: 76.6 h; minimum value: 36.1 h)
  - Feeder: 19.6 h (maximum value: 28.4 h; minimum value: 13.0 h)
- Entities average waiting time in the system:
  - Panamax: 20.0 h (maximum value: 38.4 h; minimum value: 0.5 h)
  - Feeder: 6.3 h (maximum value: 14.4 h; minimum value: 0.0 h)
- Process waiting time:
  - Berthing:
    - Berth A (panamax): 21.4 h
    - Berth B (panamax): 23.4 h
    - Berth C (panamax): 16.2 h
    - Berth D (feeder): 6.9 h

  This is the average waiting time for panamax ships and feeder ship before being served at the respective berths. In reference to berthing operations, it is the average waiting time of one ship from starting to berth until actually occupying the berth. As regards loading operations, the average waiting time of a ship before starting to load is affected by the quay rearrangement time and by the delay in the storage yard; however, as compared to berth waiting times it may be considered of no influence.

- Entities number waiting:
  - Berth A (panamax): 0.65
  - Berth B (panamax): 0.69
  - Berth C (panamax): 0.38
  - Berth D (feeder): 0.6

  This is the average number of panamax and feeder ships waiting at the port before being allowed access to the berths.
- Resources are reported in Table 3.4

Table 3.4 Number of times the quay cranes and the quays have been used at each berth, and the associated utilization percentage

Resources		Number Times Used	Utilization
Berth A	Quay A	4	86%
panamax	QC1	4	83%
	QC2	4	83%
	QC3	4	83%
	QC4	4	83%
Berth B	Quay B	5	100%
panamax	QC5	5	97%
	QC6	5	97%
	QC7	5	97%
	QC8	5	97%
Berth C	Quay C	4	71%
panamax	QC9	4	69%
	QC10	4	69%
	QC11	4	69%
	QC12	4	69%
Berth D	Quay D	13	100%
feeder	QC13	13	96%

Clearly berth D, dedicated to feeder ships, is always busy, because it is constantly used by ships over time; the only range of time the QC at berth D can rest is when the next feeder ship is berthing after the loading operations on the previous ship have been completed. With respect to the berths of the panamax ships, the number of times they are used and the associated utilization is strictly dependent on the way ships are sent to the berths (see queue conditions).

3.1.3 Simulation Results Analysis

The simulation results obtained are used to calculate the QC rate, the throughput measure of the quay cranes to estimate the simulated port productivity. Nevertheless, it is necessary to discuss first the state of the system when the simulation is over. According to port-productivity results, there are some entities still in the system when the simulation is completed, some of them processing (loading or unloading) operations at the berths, while others are in the queue waiting for accessing a berth. This means that it has not been possible to serve all the ships generated in the simulation. If the problem goal was to sell off all the entities generated, i.e., the whole demand, we could conclude that there are not enough resources to process all arriving entities: the structure of the model could have been changed, for example, by adding another berth (increasing the resources) to serve all the entities remaining in the queue, that is in order to serve all the ships arriving at the port within the simulation length. However, port operation simulation has been conducted from an operational point of view and not a planning one: thus the results at the end of the simulation can be seen as a period of transition between the state of a terminal system and a subsequent one during an unlimited temporal horizon. According to the previous considerations, the results obtained are used to illustrate how to analyze port productivity during its operation, and to see where it is possible to operate to improve terminal performance without planning a new model structure.

As mentioned above, in order to rate the terminal productivity we use the performance indicator QC rate. For convenience let us denote the numerator of QC with C (the total number of handling containers) and the denominator of QC with HQCtot (the total number of hours worked by all the quay cranes during a period). As mentioned above, quay cranes are a key resource to serve vessels in a terminal, therefore we consider resource percentage usage (see Table 3.4) especially when referring to quay-crane results, as summarized in Table 3.5.

Table 3.5 Quay-crane utilization

Quay Crane		Utilization
Berth A panamax	QC1	83%
	QC2	83%
	QC3	83%
	QC4	83%
Berth B panamax	QC5	97%
	QC6	97%
	QC7	97%
	QC8	97%
Berth C panamax	QC9	69%
	QC10	69%
	QC11	69%
	QC12	69%
Berth D feeder	QC13	96%

Multiplying each quay-crane utilization by the simulation replication length, that is 168 h, the HQCi measure (i.e., the number of hours worked for each quay crane in the simulation horizon) is obtained: for example, HQC1 is the number of hours that QC1 has worked during the whole simulation length (168 h), and so forth for all the other quay cranes. Table 3.6 reports the calculation of the HQCi for all the quay cranes.

Table 3.6 Quay cranes HQCi measures

Quay Crane		HQCi (hours)
Berth A	QC1	139.44
	QC2	139.44
	QC3	139.44
	QC4	139.44
Berth B	QC5	162.96
	QC6	162.96
	QC7	162.96
	QC8	162.96
Berth C	QC9	115.92
	QC10	115.92
	QC11	115.92
	QC12	115.92
Berth D	QC13	161.28

Thus, by calculating the total number of hours that all quay cranes have worked during the simulation, i.e., the denominator of the QC rate, we will then sum up all the HQCi values calculated above obtaining the measure HQCtot=1834.56 h.

To calculate the C value, that is the total number of handling containers during the simulation, we will consider separately unloaded containers and loaded containers. Unloaded containers are those that arrive with panamax ships and after unloading operations are sent to the storage yard. An opportunely defined variable counts each freight of these containers unloaded at each berth dedicated to panamax ships: 22,000 TEU is the total number of unloaded containers.

Loaded containers, on the other hand, are those that are charged into panamax ships, previously discharged, and into feeder ships, that arrive empty at the port for a rapid feeder service. Basically, loaded containers are all containers that have left the port in ships, therefore it is possible to know how many freights have been loaded by noting how many vessels have left the port: entities–number out simulator results show that 12 feeder ships (with a capacity of 500 TEU per ship) have left the system and that 10 panamax ships (with a capacity of 2000 TEU per ship) have also left the system; therefore, 6000 TEU have been loaded on feeder ships and 20,000 TEU have been loaded on panamax ships; 26,000 TEU is the total number of loaded containers. Thus, by calculating the total number of handling containers during the simulation, i.e., the numerator of the QC rate, we will then sum up the total number of loaded and unloaded containers obtaining M = 22,000 TEU + 26,000 TEU = 48,000 TEU.

With the above found measure it is now possible to calculate the key performance indicator and the value is the QC rate = M / HQCtot = 26.16 TEU/h that estimates the simulated port productivity.

The credibility of the results in terms of productivity is attested by comparing them with some Mediterranean Italian lesser ports (source: Confetra, 2002, www.confetra.com): in 2001 the terminals of the ports of Salerno, Venice, Trieste and Taranto valued, respectively, at 321,000 TEU, 246,000 TEU, 210,000 TEU, 186,000 TEU. Calculating these values in containers handled per hour the results are, respectively, about 37 TEU/h, 28 TEU/h, 24 TEU/h, 21 TEU/h, which are comparable to the productivity of the simulated port. Moreover, berth-allocation problem simulation shows how it is possible to increase performance in a port and keep ship turnaround time short by investing in acquisition of an adequate number of, and more efficient, quay cranes: it is to be considered that on average, a QC in Hong Kong can handle 30 containers per hour.

3.2 Final Remarks and Perspectives on Multi-objective Scenarios

We showed a port-operation simulation to calculate its efficiency. In particular, the objective discussed is that of minimizing the total time of the ships in the port. However, there is also another objective function that could be taken into account, that is the minimization of the cost of moving the containers within the port. This objective is defined as the *connectivity cost*. In a transshipment port, such as the one considered in the simulation proposed, a considerable portion of the port resources are used in transporting containers among vessels. In such a case, it is preferable that the vessels that have a large number of containers to be transported between them be berthed as close to each other as possible (see also the berth management project site, www.tliap.nus.edu.sg/tliap/Project_BerthManagement/default.aspx, for details).

The connectivity c_{ij} is then defined as the number of containers that need to be transported between any two vessels i and j. Therefore, the objective is to minimize the function $c_{ij} \cdot d(x_i, x_j)$, where $d(x_i, x_j)$ is the distance between vessels i and j, based on their berthing location x_i and x_j within the terminal.

Note that the two objectives, i.e., the minimum waiting time and the minimum connectivity cost, can be clearly in conflict, defining a complex trade-off for the port operator. This situation opens up a new scenario for the decision maker that can use simulation as a preliminary step towards a mathematical formulation approach where the problem can be modelled in terms of multi-objective optimization and solved by proper techniques able to construct, partially or completely, the Pareto-optimal front.

A similar situation occurs if the containers have to be moved from the berth location and a storage area in the terminal. Clearly, in this case c_{ij} has to be interpreted as the number of containers moved from the terminal area i (e.g., the berth location) and the terminal area j (e.g., the storage area), while $d(x_i, x_j)$ can be written as $d(i, j)$ being i and j the two terminal areas. In the following section, we will discuss a multi-objective scenario in a maritime terminal related to a container-allocation problem concerning the minimization of the distance by the containers to carry out storage and customs operations. Results will be given for a real-life case study.

3.3 Container Allocation in a Maritime Terminal and Scheduling of Inspection Operations

The goal of this section is to show a multi-objective model for optimizing, inside the customs operations, the inspection procedures for their clearance.

When a ship arrives at a port, it has to unload a certain number of containers. Among the containers that are unloaded and need customs operations, there is a fraction of them that should be inspected. The set of containers (to be and not to be inspected) are located in the terminal waiting to be delivered; next, those needing inspections, are moved from the terminal area where they have been stored in a different area where they will be processed to this end. The problem can be represented as reported in Fig. 3.2.

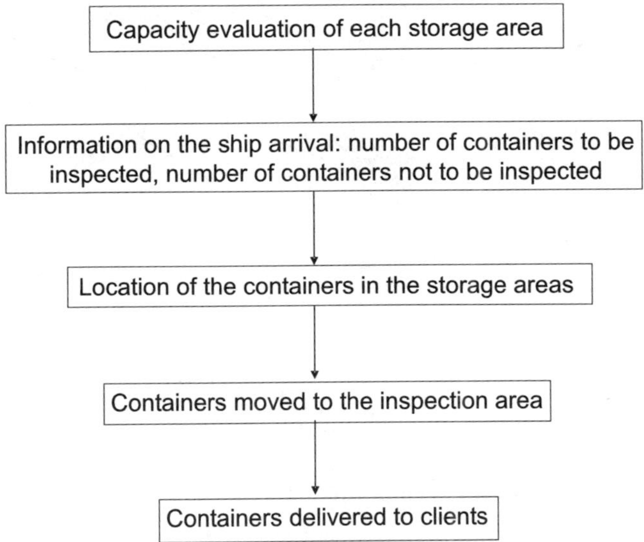

Fig. 3.2 Problem steps representation

This problem can be hierarchically divided into two subproblems: the first one is the *container allocation* in the terminal, and the second one is the *customs inspection scheduling* problem.

3.3.1 Containers Allocation in a Maritime Terminal

In the container-allocation problem one aims at allocating containers in the terminal area with the objective of minimizing the total distance covered by the containers from the ship to the terminal area they are assigned. The problem can be represented by means of an acyclic directed graph, as reported in Fig. 3.3, whose nodes represent ships and storage areas in the terminal, while arcs model the links among nodes onto which the containers flow.

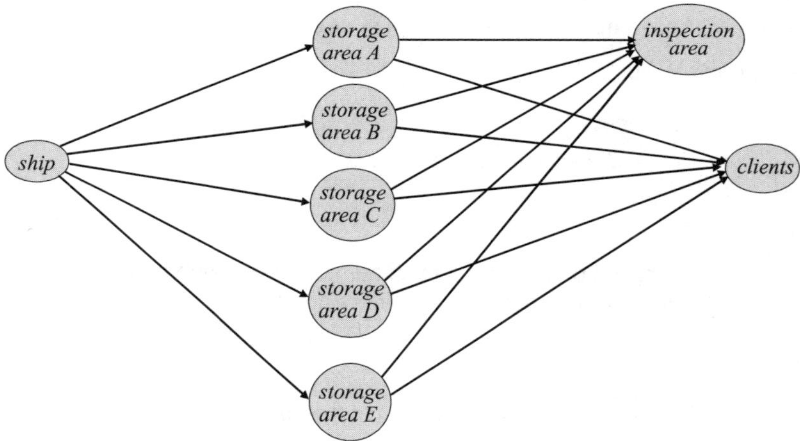

Fig. 3.3 Basic graph representation of the terminal

We add to this basic structure the variability related to the container inspection activity by modifying the above graph as follows. Assume that at time $t = 0$ a ship k arrives in the harbor, is assigned a berth and has to unload a certain number of containers. For each ship k, I_k denotes the set of containers that must be inspected, and NI_k the set of containers that will not be inspected. The inspection activities can be represented by adding nodes I_k and NI_k in the graph in Fig. 3.3, as represented in Fig. 3.4 for a generic ship k.

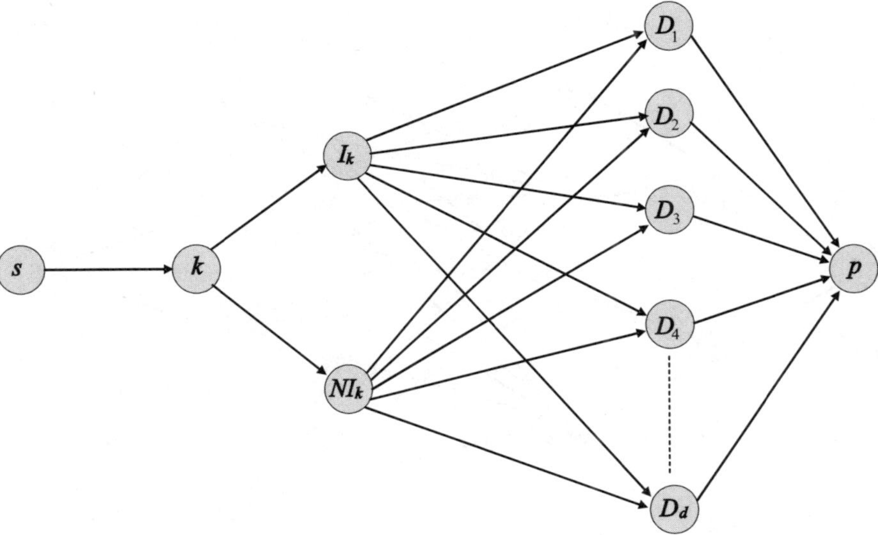

Fig. 3.4 Graph representation for a ship k

The graph in Fig. 3.4 can be easily generalized, as reported in Fig. 3.5 for two ships, i.e., nodes 1 and 2, with the following notation:

- s is the source node from which the total incoming flow of containers emanates;
- k is a ship, with $k = 1,\ldots,n$;
- I_k is the set of containers associated with ship k that have to be inspected, with $k = 1,\ldots,n$;
- NI_k is the set of containers associated with ship k that will not be inspected, with $k = 1,\ldots,n$; item D_j is a terminal area, with $j = 1,\ldots,d$;
- p is the sink node receiving the total container flow from the depots.

The problem of optimally allocating containers can then be formulated as a *minimum cost flow problem* (see also Chap. 4) on the described graph, by assigning appropriate arc costs as follows. The cost is set to 0 on the arcs joining ship nodes k to nodes I_k and NI_k. The cost is equal to the distance from the berth assigned to ship k to the storage areas D_j, for all the arcs from nodes NI_k to nodes D_j. The cost is equal to the sum of the distance from the berth assigned to ship k to the storage areas, and the distance from the latter to the inspection area, for arcs linking nodes I_k and D_j.

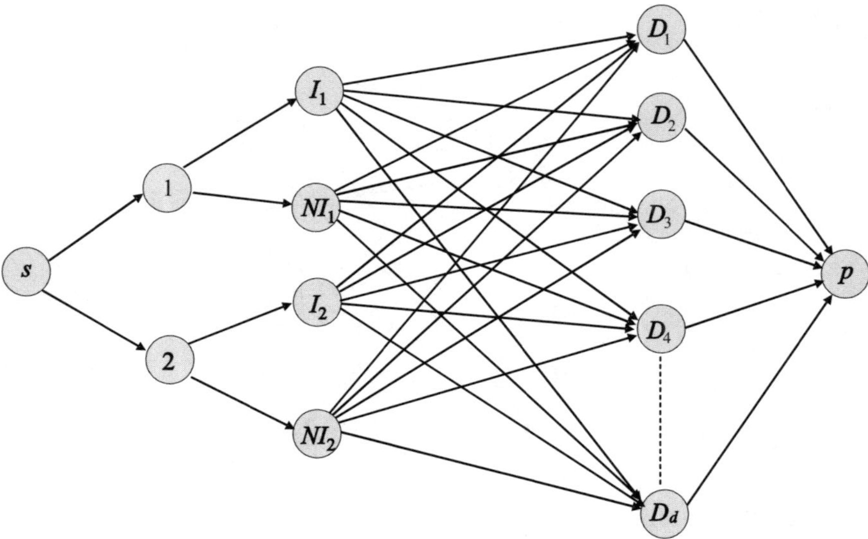

Fig. 3.5 Graph representation when $k = 2$

3.3.2 Formulation of the Allocation Model

Assume that ship k is arriving at the port and is assigned to berth b. All the containers, to be inspected and not to be inspected, must be unloaded in the different terminal areas D_j, with $j = 1, \dots, d$. The goal is to find the allocation of the containers to the terminal areas so that the transportation cost, represented here by the distances from the berth to the storage areas and from the storage areas to the inspection area, are minimized. The optimal allocation has to respect capacity and conservation flow constraints, that is:

- each berth can be assigned to at most one ship at a time;
- the number of containers that can be assigned to the terminal cannot be greater than the capacity of that terminal;
- once a ship is assigned to a berth, all its containers are assigned to that berth as well.

Let us formalize the model (at a certain time t), in a port with m berths and d terminal areas (where c_d is the capacity of terminal area D_j, with $j = 1, \dots, d$), in which n ships arrive.

$$\min \sum_{b=1}^{m} \sum_{j=1}^{d} (w_{I_b j} x_{I_b j} + w_{NI_b j} x_{NI_b j}) + \sum_{k=1}^{n} \delta s_k$$

$$\sum_{j=1}^{d} x_{I_b j} = \sum_{k=1}^{n} I_k y_{kb}, \quad b = 1, \ldots, m \tag{3.1}$$

$$\sum_{j=1}^{d} x_{NI_b j} = \sum_{k=1}^{n} NI_k y_{kb}, \quad b = 1, \ldots, m \tag{3.2}$$

$$\sum_{k=1}^{n} y_{kb} \leq 1, \quad b = 1, \ldots, m \tag{3.3}$$

$$\sum_{b=1}^{m} y_{kb} + s_k = 1, \quad k = 1, \ldots, n \tag{3.4}$$

$$\sum_{b=1}^{m} (x_{I_b j} + x_{NI_b j}) \leq c_j, \quad j = 1, \ldots, d \tag{3.5}$$

$$x_{ij} \geq 0, \quad \forall i \in \{I_b, NI_b\}, j = 1, \ldots, d$$

$$y_{kb} \in \{0, 1\}, \quad k = 1, \ldots, n, b = 1, \ldots, m$$

$$s_k \in \{0, 1\}, \quad k = 1, \ldots, n$$

The variables are:

- $x_{I_b j}$, which represents the flow of containers not requiring inspection from berth b to terminal area $j = 1, \ldots, d$, with $x_{I_b j} \in Z^+$;
- $x_{NI_b j}$, which represents the flow of containers requiring inspection from berth b to terminal area $j = 1, \ldots, d$, with $x_{NI_b j} \in Z^+$;
- y_{kb}, a binary variable equal to 1 if ship k is assigned to berth b, and is 0 otherwise;
- s_k, a binary variable, that holds 1 if ship k cannot be berthed, because all the berths are busy, and has to wait; it holds 0 otherwise.

The objective function has three components. The first two pieces are related to the distance covered by the containers in the terminal. The weight $w_{NI_b j}$ is defined as follows: it is equal to the distance between dock b and terminal area j if the container does not undergo inspection, that is:

$$w_{NI_b j} = d(b, j), \; b = 1, \ldots, m, \; j = 1, \ldots, d.$$

The weight $w_{I_b j}$ is defined as follows: it is the sum of the distances between berth b and terminal area j, and terminal area j and inspection area l if the containers must be inspected, that is:

$$w_{I_b j} = d(b, j) + d(j, l),\ b = 1, \ldots, m,\ j = 1, \ldots, d,\ l = \text{inspection area}.$$

The third part of the objective function takes into account the waiting ships. Note that in the objective function s_k is associated with a parameter δ that is arbitrarily high.

Constraints (3.1) assure that all the containers to be inspected arrived with ship k berthed in b, that is when $y_{kb} = 1$, are allocated. Similarly, we have constraints (3.2) for containers that do not need inspection. Constraints (3.3)–(3.5) assign ships to berths. Constraints (3.3) define the berth capacity: each berth can be assigned at most one ship at a time. Similarly, constraints (3.4) assign each ship to one and only one dock. Finally, constraints (3.5) ensure that the number of containers that are in a terminal area cannot exceed its capacity.

3.4 Scheduling of Customs Inspections

Let us now examine the problem of scheduling the customs inspections. Each inspection operation on a container requires a given processing time. We assume that for each container both the ready time, that is its arrival time, and its delivery time after the inspection are known. The time horizon is discretized into T periods, where a period t, with $t = 1, \ldots, T$, is a working day.

On each day a fixed number of inspections can be carried out; therefore, given the arrival time in the port and the delivery time of each container, the goal is to optimize the number of daily inspections. Note that the containers that must be delivered at time $t = 1$ must necessarily be inspected at time $t = 1$, while those that have to be delivered at time $t = 2$ can be inspected either at time $t = 1$ or $t = 2$, based on the resource availability in the inspection area; the same holds at time $t = 3$, and so forth. We represent this scenario in Fig. 3.6.

We denote with:

- o, the origin node; it represents the starting point for the inspection;
- d, the destination node; it represents the containers leaving the terminal after the inspection;
- τ_1, the node representing the inspection operation at time $\tau_1 = 1, \ldots, T$;
- τ_2, the node representing the delivery at time $\tau_2 = 1, \ldots, T$.

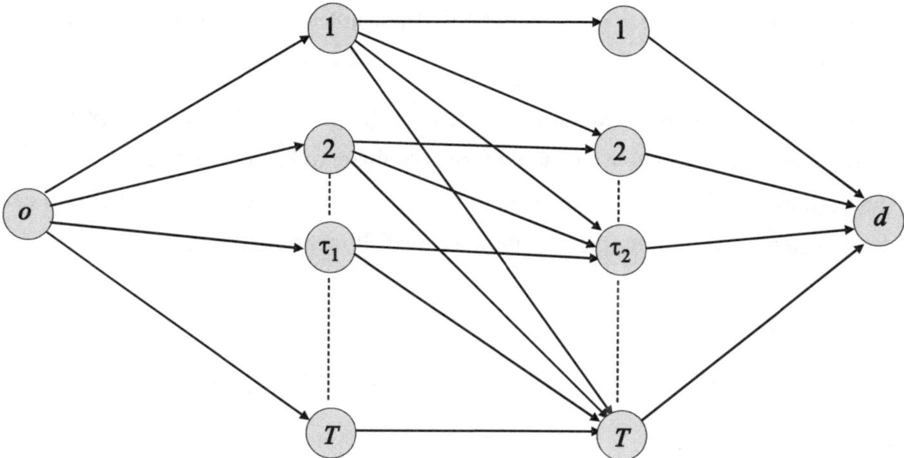

Fig. 3.6 Graph related to the customs inspections

It has to be noted that, according to the time period in which the operation is performed, the cost that will be associated to each container will change to model the delay with respect to the delivery date.

Indeed, if a container that must be delivered at a certain time t, is inspected at the same time, its cost in the terminal does not change, since once inspected it is ready to be delivered. A different analysis holds for those containers that are inspected in a time different from that of the delivery time; this is due to the fact that this container once inspected must be routed back in the terminal with a consequent additional cost and time.

To define the model, it is necessary to introduce some constraints and further notation.

- n_t, that is the maximum number of inspection operations that can be performed at time $t = 1, \ldots, T$;
- for each time $t = 1, \ldots, T$, only containers to be delivered at time $t, t+1, \ldots, T$ can be inspected;
- containers to be delivered at time t must be inspected at time $t = 1, \ldots, t$.

For this scheduling model we consider two objectives:

1. minimize the number of containers that have to be inspected at time t but that have to be delivered at a time greater than t;

2. minimize the number of containers that are not inspected before the delivery time.

Let x_t be the decision variable representing the number of container inspections performed at time t, and n_t the number of containers to be delivered at time t. In the first criterion, we are in the situation in which $x_t > n_t$; in this case, the problem can be solved by determining the minimum of the following function:

$$\min(x_1 - n_1) + (x_1 + x_2 - n_1 - n_2) + \ldots + (x_1 + \ldots + x_T - n_1 - \ldots - n_T)$$

$$= \min T x_1 + (T-1)x_2 + \ldots + x_T - T n_1 - (T-1)n_2 - \ldots - n_T$$

$$= \min \sum_{t=1}^{T} (T - t + 1)x_t.$$

In the second criterion, we are in the situation in which $x_t < n_t$; here, the objective of the problem can be formulated as follows:

$$\min(n_1 - x_1) + (n_1 + n_2 - x_1 - x_2) + \ldots + (n_1 + \ldots + n_T - x_1 - \ldots - x_T)$$

$$= \min T n_1 + (T-1)n_2 + \ldots + n_T - T x_1 - (T-1)x_2 - \ldots - x_T$$

$$= \min \sum_{t=1}^{T} -(T - t + 1)x_t.$$

Assuming that the inspection operations start at time $t = 1$, the constraints of the scheduling model are the following.

$$\sum_{t=1}^{\tau}(x_t - v_t + z_t) = \sum_{t=1}^{\tau} n_t, \quad \tau = 1, \ldots, T \tag{3.6}$$

$$x_t \leq c_{isp}^t, \quad t = \ldots, T \tag{3.7}$$

$$x_t \geq 0, \quad t = 1, \ldots, T$$

$$n_t \geq 0, \quad t = 1, \ldots, T$$

$$v_t \geq 0, \quad t = 1, \ldots, T$$

$$z_t \geq 0, \quad t = 1, \ldots, T$$

where c_{isp}^t denotes the maximum number of inspection operations that can be performed at time t. The variables v_t and z_t are defined as follows:

- v_t is the number of containers inspected that are not delivered at time t, that is in the case

$$\sum_{t=1}^{\tau} x_t > \sum_{t=1}^{\tau} n_t.$$

- z_t is the number of containers that should be delivered at time t, but have not been inspected. That is the case

$$\sum_{t=1}^{\tau} x_t < \sum_{t=1}^{\tau} n_t.$$

Thus, the cost function modelling the first criterion can be written as:

$$\sum_{t=1}^{T} v_t,$$

while the cost related to the second criterion can be written as

$$\sum_{t=1}^{T} z_t.$$

Constraints (3.6) say that the number of containers to be inspected at a certain time τ minus the number of containers that have been inspected and that are not delivered at the same time, plus the number of containers that should be delivered at time τ and that have not been inspected, must be equal to the number of containers that have to be delivered within the same time range. Assume $\tau = 1$, then we have:

$$x_1 - v_1 + s_1 = n_1$$

$$x_1 > n_1 \rightarrow x_1 - v_1 = n_1 \rightarrow v_1 = x_1 - n_1$$

$$x_1 < n_1 \rightarrow x_1 + s_1 = n_1 \rightarrow s_1 = n_1 - x_1.$$

Constraints (3.7) ensure that the number of inspection operations respects the maximum number of daily operations that can be carried out.

Hence, the objective function for this scheduling problem can be defined as follows:

$$\min \alpha \sum_{t=1}^{T} v_t + (1-\alpha) \sum_{t=1}^{T} z_t,$$

where $\alpha \in (0,1)$.

3.5 Experimental Results

The input data for the whole problem are those related to the ships (and hence the containers) arriving in the maritime terminal, and the output of the problem will be the number of containers in each storage area, that is the flow of containers leaving the terminal. The problem described above can be solved by means of the algorithmic scheme reported in Table 3.7.

Table 3.7 Solution procedure

$t = 0$	1_STEP:	Solve: Container allocation in the terminal area
	2_STEP:	Solve: Scheduling operations
	3_STEP:	UPDATE terminal capacity
$t = 1$	1_STEP:	Solve: Container allocation in the terminal area
	2_STEP:	Solve: Scheduling operations
	3_STEP:	UPDATE terminal capacity
$t = 2$	1_STEP:	Solve: Container allocation in the terminal area
	2_STEP:	Solve: Scheduling operations
	3_STEP:	UPDATE terminal capacity
...	...	...

In allocation problems, if the terminal capacity is scarce, e.g., if there are more ships than berths or if the storage-area capacity is less than the total number of containers to be placed, at least one ship must wait, and should be considered for allocation in the next time period. In the allocation model proposed in the previous section, if at time t we have $s_k = 1$, i.e., ship k has to wait, at time $t+1$ we will set a weight $\beta >> \delta$ for such a ship k to prevent the situation where it is allocated more than once.

The model resolution has been performed by using the AMPL modelling language and the CPLEX 10 solver. For the container allocation and the inspection scheduling problems, the model implementations are reported in Appendices A and B.

Computational experiments have been conducted on data form the harbor of Civitavecchia (Italy) for the period April–July 2006 (that is a period of high flow volumes). In this period, 6724 TEU have been observed.

In the experimentation we considered two berths, denoted with $B1$ and $B2$, one terminal (terminal E) composed of 28 sectors each one of capacity equal to 28

TEU/day, and an inspection area of capacity equal to 10 TEU/day. The time period considered is one week divided into 7 days.

At the considered time $t = 0$, in the terminal, there is an amount of containers equal to 165 TEU, with 48 TEU requiring customs inspection and 117 not requiring inspection. The TEU are distributed according to what is reported in Table 3.8 and 3.9.

Table 3.8 TEU in the different terminal areas at time $t = 0$

TERMINAL	TEU
E1	6
E2	7
E3	9
E4	13
E5	6
E6	3
E7	4
E8	0
E9	7
E10	2
E11	12
E12	8
E13	13
E14	6
E15	0
E16	18
E17	5
E18	6
E19	0
E20	0
E21	3
E22	0
E23	17
E24	4
E25	10
E26	6
E27	0
E28	0
TOTAL	165

Table 3.9 TEU to be and not to be inspected in the terminal at time $t = 0$

TEU TO BE INSPECTED	53
TEU NOT TO BE INSPECTED	207
TOTAL	260

At time $0 \leq t < 1$ a ship is in the harbor and has to unload its containers. This ship is assigned to berth $B2$ and the TEU are allocated to the terminal according to the residual capacity as in Table 3.10.

Table 3.10 TEU allocation at time $t = 1$

TERMINAL TO BE INSPECTED	TEU	TERMINAL NOT TO BE INSPECTED	TEU
E1	14	E5	6
E2	13	E6	17
E3	11	E7	16
E4	7	E8	20
E5	8	E9	13
		E10	18
		E11	8
		E12	12
		E13	7
		E14	14
		E15	20
		E16	2
		E17	15
		E18	14
		E19	20
		E20	5

By Fig. 3.7, it can be noticed that the produced solution attempts to locate containers to be inspected (denoted with BI) close to the inspection area (denoted with IA) so as to minimize distances covered by the containers to the inspection area. In the figure is also reported the allocation of containers not to be inspected denoted with NBI.

The total distance covered by the containers is equal to 15.5 km with an average distance per TEU equal to 60 meters.

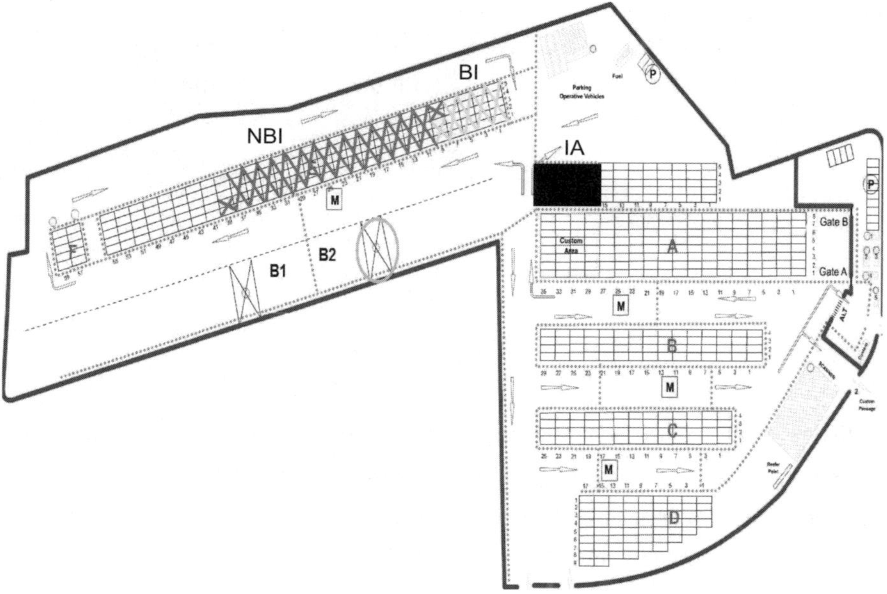

Fig. 3.7 The layout of the terminal area

Moreover, always at time $t = 1$ operations inspection must be planned as reported in Table 3.11.

Table 3.11 Planning of inspections of TEU at time $t = 1$

DAY	# OF TEU THAT MUST LEAVE THE PORT	TEU TO BE INSPECTED	TEU NOT TO BE INSPECTED
1	31	10	21
2	40	9	31
3	30	7	23
4	24	11	13
5	37	9	28
6	88	12	76
7	87	10	77
TOTAL	337	68	269

From these data, we get the organization for the inspection area reported in Table 3.12.

Table 3.12 Organization in the inspection areas

DAY	TEU INSPECTED AT DELIVERY TIME	TEU NOT INSPECTED THAT SHOULD BE DELIVERED	TEU INSPECTED BEFORE DELIVERY
1	10	-	-
2	9	-	-
3	7	-	-
4	10	4	-
5	9	-	-
6	10	6	-
7	10	-	-

In this solution it has been preferred to inspect containers exactly when they leave the terminal (values of $\alpha \leq 0.5$). If one wants to maximize the number of TEU for each time period, that is also those not leaving the terminal (values of $\alpha > 0.5$) the solution is quite different and is reported in Table 3.13.

Table 3.13 Organization in the inspection areas

DAY	TEU INSPECTED AT DELIVERY TIME	TEU NOT INSPECTED THAT SHOULD BE DELIVERED	TEU INSPECTED BEFORE DELIVERY
1	10	-	-
2	9	-	-
3	7	-	2
4	10	-	-
5	9	-	1
6	10	-	-
7	10	-	-

In this scenario, we have a greater cost, but the number of containers delivered late is 0.

Chapter 4
Hazardous Material Transportation Problems

Abstract In this chapter, the most significant results in the arena of optimal route planning and risk evaluation of hazardous material transportation will be reviewed.

4.1 Introduction

Here we are interested in presenting some general tools for transportation decision making under the assumption that safety improvement or risk minimization is an additional criterion for the decision maker that has to be balanced with distribution costs or time or other criteria.

Indeed, one should first specify who the decision maker(s) is(are), because in dealing with safety it is implicit that the inhabitants, drivers and enterprises can be seen as different stakeholders representing different interests. For instance, considering road transportation only one should take into account land, environmental and residents issues, drivers aspects, vehicle characteristics and also goods category.

Hazard analysis is the process adopted to assess the risk and, in this chapter, we assume that the risk associated with an event (e.g., the probability of the event times the consequences of the event) has been already computed. Indeed, estimating a risk requires the analysis of transportation accidents that is very complex and would involve a thorough study for each specific case that goes beyond the scope of this chapter. For the sake of completeness, here we recall just a few ideas on this topic. For example, the estimation of the probability of an accident in a road transportation is carried out by means of an *event tree* that models the cause–effect relationships entering in the final event. In this sense, the accident is seen as the occurrence of its primary causes and of intermediate concatenated events.

In the remainder of this chapter, we will present optimization models related to the minimization of the risk, concentrating on hazardous material (hazmat) transportation. The main target of the studies on hazmat transportation is to select the route(s) from a given origin o to a given destination d such that the risk for the surrounding population and the environment is minimum, without producing excessive costs for the carriers. A definition of risk will be given in the next sections; roughly speaking, we can say that the risk is a measure of the consequences associated with an accident in which a hazardous material is involved.

In order to give an idea of the impact of hazmat on the population, we show some data drawn from the 2004 report "Creating A Disaster-Resistant Lower Colorado River, Basin Texas Colorado River Floodplain Coalition, Hazard Mitigation Action Plan for the Lower Colorado River Basin 2004–2009, H2O Partners, Inc". In particular, in Tables 4.1 and 4.2 we give the hazardous materials incidents by transportation modes from 1883 through 1990 in Texas, and the number of incidents in the City of Austin from 1993 to 2001.

Table 4.1 Hazardous materials incidents by transportation mode from 1883 through 1990

Mode of Transportation	Number of Accidents	Associated Deaths	Associated Injuries
Air	1220	0	153
Highway	41,781	79	1569
Railway	7886	1	423
Water	83	1	35
Other	29	0	2
Total	50,999	81	2182

The numbers are quite explicit and suggest that tactical and operational policies to manage this kind of transportation are sought in order to guarantee safety for the population exposed.

In the arena of single-objective approaches to hazmat transportation, papers in the literature are mainly concentrated on studying techniques for selecting a (set of) paths minimizing one among the following criteria:

- the travel duration (Brogan and Cashwell, 1985);
- the expected number of deaths or injured people (Batta and Chiu, 1988);

Table 4.2 Hazardous materials incidents in the City of Austin, 1993–2001

Year	Incidents
1993	20
1994	28
1995	34
1996	32
1997	32
1998	15
1999	34
2000	21
2001	4
Total	220

- the probability of an accident with release of hazardous material (Saccomanno and Chan, 1985);
- the risk exposure of the population resident along the selected routes (Glickman, 1983);
- the risk of accidental release of hazardous material (Patel and Horowitz, 1994).

Before going in depth in the analysis of the state-of-the-art of hazmat transportation with multiple objective functions, it is worthwhile to summarize the results of the 1985 article of Saccomanno and Chan. Here, the authors proposed an interesting empirical cost–benefits analysis (based on data related to the City of Toronto) of three different optimal routing strategies for hazmat transportation: the first consists in minimizing the transportation costs; the second in minimizing the accident probability, and the third in minimizing the risk exposure of the population resident along the chosen routes. After the evaluation of the impact of random variations of some environmental factors on the selected routes (for example, the road conditions and the visibility), they achieved the following conclusions:

- the strategy that minimizes transportation costs tends to privilege the economic advantage instead of the social security;
- the strategies that minimize either the incident probability or the risk of the exposed population tend to privilege the social security instead of the economic advantage;

- none of the three strategies considered is able to generate an "absolutely safe" route, since random variations of environmental factors can have unexpected effects on security;
- the best strategy to adopt, from an economic point of view, is to minimize the risk; even though in this case the transportation cost can increase significantly, the social savings resulting by a smaller number of incidents are significantly greater.

Notwithstanding the interest of these results, single-objective models for hazmat transportation are not able to represent the conflicts arising when more than one criterion is taken into account, e.g., in the case of the transportation cost and the risk minimization. Indeed, in the latter case, routes associated with a low risk tend to be very expensive because of their possible larger length. This is why many authors recognize the intrinsic multi-objective nature of hazmat transportation (with conflicting objectives) as explained in the next section.

4.2 Multi-objective Approaches to Hazmat Transportation

Kalelkar and Brooks (1978) and Shobrys (1981) work represent the first attempt to treat explicitly the multiple objectives framework in hazmat transportation.

Kalelkar and Brooks described an approach based on decision theory for selecting the better route from a Pareto-optimal route set, given generic multiple conflicting objectives. Shobrys considered in particular two objectives: the minimization of the route length and the amount of the population exposed to risk and located in a certain range of the selected route. For different combinations of the two objectives by using various weights, he first assigned a hybrid distance–population cost to each arc of the transportation network; next, he solved the shortest-path problem for obtaining a set of Pareto-optimal solutions.

A similar scalarization technique was considered in McCord and Leu (1995). In this work, the authors studied a bi-objective problem with the minimization of the transportation cost and the exposure to the risk of the population. They resorted to multi-attribute utility theory transforming a multi-objective problem into a single-objective one; in this way, as for the previous case, a shortest-path algorithm was used for determining the best path with respect to the scalarized objective function.

McCord and Leu observed, in particular, that the selected routes are very sensitive to small changes of the utility function parameters, and, therefore, by varying such parameters, it is possible to determine a significant Pareto-optimal set of routes.

Current et al. (1988) described a method for selecting Pareto-optimal paths minimizing the route length and the population coverage obtaining a minimum covering shortest-path problem.

However, as we described in the chapter devoted to multi-objective optimization, solution methods based on exploiting a linear combination of the objectives are not able, in general, to determine the whole set of non-dominated solutions (Pareto-optimal solutions).

4.2.1 The Problem of the Risk Equity

When many vehicles have to be routed among the same origin–destination pair, a portion of the population (or a particular geographic area) could be possibly exposed to a high risk if all such vehicles were routed on the same path. In this case, there is the need to distribute the risk in an equitable way over the population. Equity can be clearly interpreted as a further objective to be attained.

Zografos and Davis (1989) studied a procedure to solve the multi-objective shortest-path problem determining a set of routes minimizing simultaneously

1. the total risk,
2. the risk for particular categories of population (for example, those located in hard-to-evacuate areas),
3. the travel time,
4. the damages of properties,
5. the risk equity.

Gopalan et al. (1990a,b) presented an integer programming formulation for the problem of minimizing the total risk associated with a set of k origin-destination routes, able to guarantee equity. The latter is achieved by constraining the differences of the risks associated with every pair of regions in which the geographical area is divided, to be less than or equal to a given threshold.

Similarly, Lindner et al. (1990) tried to determine a set of routes distributing the risk in an equitable way over time between every pair of geographic regions, when several shipments have to be carried out in a certain time interval among the same origin–destination pair. By using the model for measuring the risk along a route described by Batta and Chiu (1988), the authors first formulated the problem in terms of integer programming and then in terms of dynamic programming.

Also, List and Mirchandani (1991) formulated a multi-objective problem minimizing risk, cost, and equity for which the Pareto-optimal solutions set is given.

Finally, we mention the paper of Jin et al. (1996) where the authors showed that the set of origin–destination pairs for which the risk equity is attained is in general associated with a higher total risk than the set for which the risk is distributed in a less equitable way.

4.2.2 The Uncertainty in Hazmat Transportation

Many authors recognized the uncertain nature of the risk and/or of other characteristics of hazmat transportation: the number of incidents on a road, the amount of population living and/or working near a route, the travel duration, the effects of the release of a hazardous material, can be modelled by means of random variables whose distributions may vary over time.

However, on the one hand, variable distributions are hard to determine in many cases due to the scarcity of data, and, on the other hand, the probability to be estimated, have a very small value (they are often associated with events that perhaps never happened).

Turnquist (1993) determined the Pareto-optimal route set with respect to the following objectives: minimization of incident rates related to the release of hazardous material; minimization of the population exposed to the risk and minimization of the route length. The incident rates and the population exposure were assumed probabilistic and considered unchangeable over time.

Also, Wijrante et al. (1993) dealt with a multiple objective problem with uncertain attributes (unchangeable over time) and presented two methods, based on an approximation of the stochastic dominance, for comparing the distribution functions of a single stochastic criterion on each path. These methods are then extended to include multiple stochastic criteria, which are in turn reduced to two deterministic criteria; therefore, the final problem is a multi-objective deterministic problem.

4.2.3 Some Particular Factors Influencing Hazmat Transportation

Some other factors can influence the selection of routes for hazmat transportation.

Cox and Turnquist (1986) described optimal (scheduling) strategies for determining the best departure times to transport hazardous materials on fixed routes traversing cities where heavy vehicles are allowed only in certain time windows. In this case, the travel time is modelled both as deterministic and stochastic. The authors observed that, since the delay caused by time-window constraints tends to increase the population exposure time to the risk of hazmat release, the optimal departure time is the one that produces the minimum delay.

Other authors observed that the decisions related to the public security should not be based on the "expected value", since the latter is not able to model properly "low probability–high consequence" scenarios.

Glickman (1983) considered the relationships between the expected and the worst-case risk value in deciding alternative routes for hazmat transportation.

Sivakumar and Batta (1994) formulated and solved a variance-constrained shortest-path problem. This model was applied to find a safe route for GPL transport, that can cause considerable damages (its dispersion in the air is highly unforeseeable).

Boffey and Karzakis (1995) demonstrated that non-linear models minimizing the risk are the most appropriate in high-risk scenarios.

4.2.4 Technology in Hazmat Transportation

Recent developments in the hazmat transportation area are represented by the technologies providing a support either in the route-selection phase (see, for example, geographical information systems, GIS) or in the selected route-control phase (see, e.g., automatic location systems of a vehicle like geographical position systems, GPS, or automatic vehicle location, AVI).

The main GIS characteristic is to treat a large amount of different information in a unique visual ambient. In particular, they can combine, in a unique visual representation, data related to the transport network structure, and the demographical, economical and geographical characteristics of the zones involved by the selected routes.

Lepofsky and Abkowitz (1993) discussed the benefits of using GIST, i.e., GIS for transportation, (in the hazmat context) that can help in the risk evaluation, in routing and scheduling, in preparing evacuation programs, and in the general management of the emergencies.

Beroggi and Wallace (1995) developed models for selecting alternative routes for vehicles in hazmat transportation, when an incident happens on their initial route. They demonstrated that advanced communication technologies, integrated with a good decision-support system, can lead to better choices and, hence, to safer transports.

However, it is worth noting that, as emphasized by Miller and Mahamassani (1998), even if advanced technologies are employed, both risk and travel time should always be considered as variable since the network characteristics tend to vary over time. To this end, Miller and Mahamassani in their work posed the target to find paths minimizing both risk and travel time, assuming the latter as stochastic variables with density distribution varying over time.

4.3 Risk Evaluation in Hazmat Transportation

4.3.1 Risk Models

As we previously said, the difference between the hazmat transportation and any other kind of transport is mainly the risk (for people and environment) associated with the release of the transported material in the case of an incident.

To define the risk, it is necessary first to distinguish the risk in the following classes:

- the risk for the population;
- the risk for the environment.

The risk for the population in the literature is divided into three subclasses:

- *the individual (or local) risk*, which refers to the probability that a human being, without any kind of protection and located in a fixed point in the impact zone, undergoes a considerable injury in a given interval of time (in general one year);
- *the social risk*, which refers to the number of expected deaths in a given interval of time (in general one year), or to the probability that, in the same interval of

time, an incident with a number of deaths greater than a certain threshold happens;

- *the collectively perceived risk*, which measures the risk aversion by the public and the authorities.

These kinds of risk can be specialized in:

- risk related to the used routes;
- risk related to the transported material;
- risk related to the transport mode;
- risk related to the environmental conditions (e.g., weather, road);
- risk referred to other factors different from the previous ones.

The points reported in the last classification can overlap each other.

In the following, we will refer mainly to the social risk related to the used routes. In particular, in the next paragraphs, we will start analyzing the traditional risk model with its simplifying hypothesis; next, other risk models will be discussed. Finally, we will review an axiomatic approach to the risk definition. The analysis refers to the Erkut and Verter (1998) paper.

4.3.2 The Traditional Definition of Risk

The first definition of risk used in the hazmat transportation literature is that related to the social risk. The social risk can be defined as the product between the incident probability and its consequences (accidental number of deaths).

Let $p_{A,B}$ be the incident probability on the road segment A when the material B is transported and let $C_{A,B}$ be the population resident along A inside the impact zone of the transported material B. The risk associated with moving material B along A can be defined as:

$$R_{A,B} = p_{A,B}C_{A,B}.$$

It is assumed that the arcs of the graph representing the road network have been defined in such a way that the incident probability and the density of the population resident on each arc have a uniform distribution. For example, a long road traversing both the city center areas and the city suburbs, should be represented by a series of arcs, each one with an incident probability and a population density that can be assumed to be distributed uniformly.

Therefore, a generic arc of the road network can be modelled as a sequence of "unitary segments"; neglecting the indexes related to road type and the transported material, the risk on an arc formed by n unitary segments is:

$$R = pC + (1-p)pC + (1-p)^2 pC + \ldots + (1-p)^{n-1} pC.$$

Due to the uniform distribution hypothesis, each unitary segment of an arc has the same incident probability p and the same population quantity C.

An incident can happen in the first segment with probability p and causes the consequence C, or it can happen in the second segment with probability $(1-p)p$ and causes the consequence C, and so on until the nth segment.

If we take into account that $p^s \approx 0$, with $s > 1$, the risk definition can be reduced to a simple expression that defines the risk as an attribute of the single network road segment:

$$R = npC.$$

Generally speaking, the risk R_i associated with arc i is:

$$R_i = p_i C_i,$$

where p_i is the incident probability on arc i and C_i is the population inside the impact zone near arc i.

The above definition implicitly assumes that each individual resident in the impact zone is subject to the same undesirable consequence. In practice we know that the consequence of the release of a hazardous material in the environment depends on several factors: the weather conditions, the time of the day, the territory topography, the distance from the point at which it happens, the response of the human organism, and so on. However, all these information are hard to obtain, even if they exist in theory.

Let P be a generic path with origin o and destination d formed by m arcs, and let $R(P)$ be its total risk; we can write:

$$R(P) = p_1 C_1 + (1-p_1) p_2 C_2 + \ldots + (1-p_{m-1}) p_m C_m.$$

An incident can happen in the first, in the second or in the mth arc of the considered route P with consequences C_1, C_2,...,C_m, respectively.

Assuming $p_i = p_j \approx 0, \forall i, j$, for the same reason as that $p^s = 0, s > 1$:

$$R(P) = p_1C_1 + p_2C_2 + \ldots + p_mC_m.$$

The total social risk of route P between origin o and destination d is approximately equal to the risk sum of the arcs constituting p. Therefore, the minimum-risk route-selection problem can be reduced to a shortest-path problem.

The interested reader will find in the Erkut and Verter (1998) work observations about the correctness of the approximation $p_i = p_j \approx 0, \forall i, j$.

4.3.3 Alternative Definition of Risk

Even if the risk definition given has a general validity, it can be considered completely inadequate in some particular scenarios. To this aim, in the literature, it is possible to distinguish at least four other risk models:

- the incident probability $IP(P) = \sum_{i \in P} p_i$ (Saccomanno and Chan, 1985, Abkowitz et al., 1992);
- the total exposed population $EP(P) = \sum_{i \in P} T_i$, where $T_i = C_i d_i$, $i \in P$, and d_i is the length of arc i (ReVelle et al., 1991, Batta and Chiu, 1988)
- the perceived risk $PR(P) = \sum_{i \in P} p_i (C_i)^q$ (Abkowitz et al., 1992);
- the conditional risk $CR(P) = \frac{\sum_{i \in P} p_i C_i}{\sum_{i \in P} p_i}$ (Sivakumar et al., 1995).

The concept of exposed population (total number of people exposed to the risk) can be applied when the transported hazardous material tends to generate a risk for the population near the route in terms of exposure rather than incident.

In this scenario, referring to the social-risk definition given, we can interpret p as the probability of exposure of people living in the neighborhood of the path; it can be assumed that $p = 1$, i.e., there is an exposure to the hazardous material, even if very small.

Although the exposure to small amounts of hazmat for short time intervals does not represent particular risks for the human health, it is reasonable to expect that the public opposition to these transports will be proportional to the exposed population size. Therefore, selecting routes minimizing the exposure of the population can be interpreted as the minimization of the opposition to hazmat transportation by public opinion.

On the other hand, let us suppose that the transported material has a very small impact radius so that C can be disregarded; in this case, the objective can be to minimize the incident probability in order to minimize the risks imposed on other motorists.

The public does not perceive in the same way the events with low probability and high consequence, and those with high probability and low consequence, even if they present the same expected consequence. For this reason the traditional concept of risk appears not to be adequate.

Abkowitz et al. (1992) suggested to represent the perceived risk (*PR*) (for a unitary segment of arc) through a preference parameter q as

$$PR = pC^q.$$

So the *PR* associated with route P is

$$PR(P) = \sum_{i \in P} p_i (C_i)^q.$$

The aversion to the consequences of a potentially risky hazmat transport can be indicated with $q > 1$. When $q > 1$ the arc weights increase proportionally to the population resident along them, and a minimum-risk path-selection algorithm will tend to select arcs traversing not very crowded areas.

Sivakumar et al. (1995) suggested a conditional-risk definition, as the expected consequence at the instant of the incident time:

$$CR(P) = \frac{\sum_{i \in P} p_i C_i}{\sum_{i \in P} p_i}.$$

It is clear that there is a fundamental difference between the conditional-risk model and the other risk models: the latter are single-objective models (they can be represented by a min-sum objective function) and can rely on standard shortest-path solution algorithms, the former is not linear and hard to solve.

For a more precise and structured comparison, in the next paragraph an axiomatic approach will be introduced together with a risk model for hazmat transportation based on these axioms.

4.3.4 An Axiomatic Approach to the Risk Definition

Erkut and Verter (1998) proposed an axiomatic approach to the risk definition, which will be partially recalled in the following.

Let $P1$ be the set of all the paths between origin o' and destination d' and $P2$ the set of all the paths between origin o and destination d. Let $P_1 \subset P_2$ for all $P_1 \in P1$ and for all $P_2 \in P2$. Let $V(P) = R(P)$ the risk function that evaluates path P. We can define the following axioms.

Axiom 1. $V(P_1) \leq V(P_2)$ for all $P_1 \in P1$ and for all $P_2 \in P2$.

This means that adding one or more arcs to a given path P, the risk of the path so obtained cannot be less than that of the initial path.

Axiom 2. $V(P_2) = \min_{P_2 \in P2} V(P)$ implies $V(P_1) = \min_{P \in P_1} V(P)$.

It establishes that all the subpaths of an optimal path must be optimal; it is the optimality principle of Bellman applied to the hazmat transportation context.

Let $V(P) = R(p(P), C(P))$, where $R(\cdot)$ is a risk function, $p(\cdot)$ is the accident probability vector on the arcs of P, and $C(\cdot)$ is the consequence vector.

Axiom 3. $p(P) \leq p'(P)$ and $C(P) \leq C'(P)$ imply $V(P) = R(p(P), C(P)) \leq R(p'(P), C'(P)) = V'(P)$.

It establishes that the path risk is a non-decreasing function of the accident probability and of the consequences associated with the arcs of the same path. In other words, the increasing of the accident probability and/or of its consequences on an arc may not involve a risk reduction on the whole path.

Models associated with traditional risk (only if the assumption $p_i p_j \approx 0$ holds), incident probability, exposed population, perceived risk, satisfy all the axioms; the conditional-risk model instead does not satisfy any of them since the denominator in its formula is equal to the incident probability on the whole path; therefore, routes with high incident probability will tend to have low conditional risk.

The fact that one or more of the axioms listed above is not verified, is not merely a theoretical issue. Suppose that, for instance, referring to traditional risk, all routes between an origin and a destination have the same expected consequence (the value of the numerator); in this case, the conditional risk model would lead to a path with higher incident probability. Moreover, as a further example, the increase of the incident probability on an arc with null population (in the case of the sabotage on that arc, for instance) would render the routes crossing this arc the most convenient ones.

The reader will find a review of very interesting paradoxical situations in Erkut and Verter (1998).

4.3.5 Quantitative Analysis of the Risk

By the analysis of the previous sections, it emerges that risk definition and evaluation is very important; however, there does not seem to be a unanimous agreement among researchers on these issues. In the following, we try to give an answer to the following question: how can the risk be calculated in practice?

Risk evaluation in hazmat transportation is somehow similar to what is done in other contexts, like risk evaluation in industrial plants. However, there are considerable difficulties in the evaluation of the accidents frequency and their consequences, mainly caused by the scarcity of historical-statistical data and the variability of the conditions along the routes under evaluation. For this reason, as said in the previous paragraphs, each route is divided, if necessary, into segments of unitary length, onto which the parameters to be evaluated can be assumed to be constant.

A general procedure for the risk evaluation can be divided into three phases:

- evaluation of the probability of an undesirable event (for example, an incident with a release of toxic-harmful substance);
- evaluation of the exposure level of the population and the environment, given the nature of the event;
- evaluation of the degree of the consequences (e.g., deaths, injured people, damages) given the exposure level.

In theory, each step of this process produces one or more probability distributions. The latter can be combined to obtain the potential consequences distribution. However, in practice, this three-step approach is not followed in all its phases; a frequent behavior is to consider only the expected value of these distributions, determining the "expected damage" as a measure of risk. In other situations, the analysis is concentrated on the second phase of the procedure, considering the population exposure as a "worst case" measure of the risk.

The estimates of the incident probabilities are generally based on:

- estimates of the incident rate concerning vehicles transporting hazardous materials;
- estimates of the material-release probability.

It is clear that, for a given application, the best would be to estimate the probability that a vehicle is involved in an incident with release of hazmat, in a given location and under given environmental conditions. However, the available data do not possess, in general, this degree of detail. Indeed, data related to incidents and release of hazardous material, instead, are generally limited to a specific time window in a day. If exposure measures are also considered (for example, km per vehicle), an estimate of the incident rate would be obtained. There are three main difficulties in determining such rates:

- selecting from the whole incidents database those that are relevant to the event under consideration;
- determining a measure of exposure consistent with such an event;
- approximating such estimates considering the low number of incidents for each category due to the lack of data available.

The risk analysis connected to a certain event is in general represented through the risk profile, that is the function

$$[1 - F(x)] = P(X > x),$$

where $F(x)$ is the cumulative distribution function of a random variable X "annual number of deaths"; so x is the consequences measure of a certain activity.

The risk profile returns the probability that the consequences of a certain event exceed a fixed threshold; moreover, it is a more detailed measure than the expected value that, instead, represents the expected number of deaths.

A review of works using these concepts in real applications is presented in List et al. (1991); Zhang et al. (2000) and Brainard et al. (1996) treated the specific problem of risk estimation in hazmat transportation. Bonvicini et al. (1998), instead, proposed to use fuzzy logic for the risk estimation. Wijeratne et al. (1993) used two parameters: the average incidents number and the variance of the normal distribution associated with this number. Miller-Hooks and Mahamassani (1998) treated the risk as a random variable and its probability distribution is assumed to be variable over time.

4.4 The Equity and the Search for Dissimilar Paths

We have previously observed that the problem of distributing the risk in an equitable way can not be neglected by routing all the hazmat flow on the optimal path with minimum cost/length.

The equity implies the search for an alternative set of routes between origin–destination pairs. A method to find routes with a fairly distributed risk consists in generating dissimilar origin–destination paths. Given a graph representing the road network, two paths joining the same pair of origin–destination nodes are called dissimilar (or alternative) if at least one arc distinguishes them.

The dissimilarity of routes is useful if it is suitably defined in spatial terms, since it allows the selection of routes avoiding the selection of the same geographical areas, hence distributing the total risk in an equitable way.

The dissimilar paths generation problem has been dealt with in the literature in at least four different ways, based on the application context. In the following, we will review these methods: iterative penalty method (IPM), gateway shortest paths (GSPs) method, minimax method (MM), *p*-dispersion method.

4.4.1 The Iterative Penalty Method

The iterative penalty method (IPM) (Johnson et al., 1992) is, among the methods discussed in this section, the only one explicitly studied for hazmat transportation. It consists essentially in the iterative application of a shortest-path algorithm. At the end of each iteration, the arcs in the current path are penalized by increasing their weights; this discourages the selection of the same arc set in the generated paths set in the next iteration, with a consequent possibly dissimilar paths set (all between the same origin–destination pair).

In this penalization mechanism:

- penalties can be assigned either to the last generated path or to all the paths generated until the current step;
- penalties can be assigned to arcs, nodes, or both;
- penalties can be applied to the arc weights, either of the current or of the initial iteration
- penalties can follow an additive mechanism (i.e., a fixed positive quantity is added to the arc weights of the current/initial iteration), or a multiplicative one (i.e., the arc weights of the current/initial iteration are multiplied by a factor $\lambda > 1$).

The results of this method clearly depend on the choice of the penalization mechanism; so, small penalties can lead to scarcely dissimilar paths, just like large penalties can exclude from the analysis a certain number of possibly "good" paths. Furthermore, for large penalties the method tends to return a lower number of paths with a large dissimilarity than the case of small penalties. Note that, in order to prevent the selection of paths that are too long, all routes whose length exceeds a fixed threshold can be discarded.

The advantage of this method lies in its simplicity: one needs only a shortest-path algorithm and a penalization mechanism. However, its simplicity is also its greater limit since the method does not offer any evaluation subroutine for the quality of the generated paths.

4.4.2 The Gateway Shortest-Paths (GSPs) Method

The GPSs method was introduced by Lombard and Church in 1993 during the study of oil pipeline localization. An oil pipeline has to be not too long, has to respect the territory topography and has to take into account the closeness to the cities it has to service.

Assume that the shortest path from origin o to destination d (along which the oil pipeline has to be built) cannot be implemented due to the above constraints. For the same reason, assume that a slightly (spatially) different path from the shortest path is not valid either. Therefore, it is necessary to generate a set of not too similar (topologically) paths with respect to the shortest path, and to choose the desirable paths in the set so found. This is the problem dealt with and solved by Lombard and Church by means of the GSPs method.

A gateway shortest path, given an origin–destination pair, is a shortest path constrained to contain an assigned node (the gateway node).

In this way, it is possible to generate a large set of dissimilar paths, simply forcing, each time, a new path to pass from a different gateway node. In order to evaluate the dissimilarity between two paths, the concept of "area under the path" is used. That is, the network is assumed to be representable on a plane so that the "area under the path" is the area between the path and the x-axis. Therefore, the dissimilarity between two paths is measured by the absolute difference between the areas under the paths.

In the hazmat transportation context, the GSPs method could be applied in the following way: given a pair of origin–destination nodes, all gateway paths are generated initially; then from this set, through the Tchebycheff metric, an efficient subset is selected based on the dissimilarity and the length of paths in the subset.

The GSPs method, as the IPM method, is easy to implement, but differently from IPM has a way of evaluating the dissimilarity among the generated paths. However, this dissimilarity is computed with respect to the shortest path and is not computed as a mutual dissimilarity among the paths, that is able to guarantee an equitable distribution of the risk.

More detailed experimental results performed by Akgün et al. (2000), show that:

- many of the initially generated paths can contain cycles, and the greatest part of routing applications in the hazmat transportation context does not prefer such situations;
- it cannot be possible to identify any dissimilar path as a gateway shortest path even if it is rather desirable;
- a desirable GSP can be identified as a dominated path and can be deleted in the final phase of the method;
- the GSPs method can be prone to generate a very similar final path set (even if very dissimilar from the shortest path) and very long.

4.4.3 The Minimax Method

The Minimax method was proposed by Kuby et al. (1997) in the context of "multi-commodity flow" problems. The target of this method is to select a certain number of origin–destination paths that are not too long and have the minimum mutual number of arcs in common. Indeed, if a very similar arc set is used by two of these origin–destination paths, the flow of commodities traversing them will result in an increase of the risk for the population resident in their neighborhood. In general, the lower the number of arcs in common among the paths, the greater is the potential total capacity of flow between the origin and the destination.

The Kuby algorithm first generates, through a k shortest-paths algorithm, a set of alternative paths from an origin o to a destination d; next, among them, it selects iteratively a subset of dissimilar paths (DP) by means of an index that determines the inclusion or not of candidate paths in DP.

The first path included in DP is the shortest path, denoted with P_1 and $d(P_1)$ its length. The second path is chosen among the k shortest paths in such a way that its length and its similarity with P_1 are minimized.

Let k-SHPs be sets of the k shortest paths and let $P_j \in k$-SHPs be a generic path. The length of P_j can be expressed by means of the distance $d(P_j, P_1)$ between P_1 and P_j. Denote with $d^s(P_j, P_1)$ the length of the common parts between P_j and P_1 (the similarity), and with $d^n(P_j, P_1)$ the length of the parts of P_j not in common with P_1; then, the distance between P_1 and P_j will be $d(P_j, P_1) = d^s(P_j, P_1) + d^n(P_j, P_1)$.

The second path to include in DP will be

$$P_j \in k - SHP, \text{ such that } M(P_j, P_1) = \min_{z \in DP} M(P_z, P_1),$$

where

$$M(P_z, P_1) = \frac{[(1+b)d^s(P_z, P_1) + d^n(P_z, P_1)]}{d(P_1)}; \quad b \geq 0.$$

The minimization of the index $M(P_z, P_1)$ with respect to z leads to a relatively short path that is also dissimilar from the shortest path. The procedure continues until the desired path number is generated.

If

$$M(P_j, P_i) = \frac{[(1+b)d^s(P_j, P_i) + d^n(P_j, P_i)]}{d(P_i)}; \quad b \geq 0$$

the Kuby model becomes equivalent to

$$\min_{j \in DP}[\max_{i \in DP} M(P_j, P_i)].$$

Minimax is the only method among those considered in this section to consider explicitly the mutual dissimilarity among all pairs of generated dissimilar paths. However, its use is not recommended in the hazmat transportation context, because of some inconveniences:

- the $M(P_j, P_i)$ index depends on b that represents the dissimilarity weight between P_i and P_j. If, on the one hand, a suitable calibration of this weight can allow one to privilege one of the two objectives (cost and dissimilarity), on the other hand, its exact evaluation should be left to the manager experience;
- the length $d^n(P_j, P_i)$ of the parts of P_j not in common with P_i is not symmetric. If P_i is not equal to P_j then $d^n(P_j, P_i)$ is not equal to $d^n(P_i, P_j)$ unless the two path lengths are equal. Due to this asymmetry, the final DP set can depend on the selection order of its members, as shown by Akgün et al. (2000) with a simple numeric example.
- the method is prone to determine not very dissimilar paths, as far as the k shortest-paths algorithm tends to generate paths not very different among them, i.e., not very different from the shortest path, unless a very large value for k is chosen with a consequent increase of the computational time.

4.4.4 The *p*-dispersion Method

With respect to the methods discussed so far, the p-dispersion method allows a better mutual dissimilarity among the generated path set, that, as we pointed out, guarantee an equatable distribution of the risk.

Let U be a set of np points and $d(i, j)$ be a measure of the distance among the points $i, j \in U$; let $S \subset U$ be a generic subset of p points $(1 < p < np)$.

If $d(P_i, P_j)$ is the dissimilarity among paths P_i and P_j, a set of dissimilar origin–destination paths may be determined as follows:

- generate an initial set U of candidate origin–destination paths;
- determine a maximal dissimilar subset S (of size p, with $1 < p < np$), i.e., the one with the maximum minimum dissimilarity among its paths.

Therefore, the p-dispersion method formulation is:

$$\max_{S \subset U} \min_{i,j \in S} d(P_i, P_j), \quad \forall i \neq j.$$

The definition of the index $d(P_i, P_j)$ is based on the parts that P_i and P_j have in common, i.e.,

$$d(P_i, P_j) = 1 - \left\{ \left[\frac{L(P_i) \cap L(P_j)}{L(P_i)} \right] + \left[\frac{L(P_i) \cap L(P_j)}{L(P_j)} \right] \right\},$$

where $L(\cdot)$ is the length of the considered paths. So $d(P_i, P_j) = 1$ means there is maximum dissimilarity, while $d(P_i, P_j) = 0$ means there is maximum similarity.

However, it can happen that two paths, without any part in common, are parallel but very close (for example 100 meters), but would be considered by this index as having maximum dissimilarity (equal to 1). It is clear that such a situation is impractical in the hazmat transportation context, because the greatest part of the materials disperse in the air inside a certain impact zone, after of an incident, producing a high risk for the population near the two paths and therefore representing a non-fair solutions in terms of risk-equity distribution.

From here, the need for a new definition of dissimilarity, i.e.,

$$d(P_i, P_j) = 1 - \frac{1}{2} \left\{ \left[\frac{A(P_i) \cap A(P_j)}{A(P_i)} \right] + \left[\frac{A(P_i) \cap A(P_j)}{A(P_j)} \right] \right\},$$

where $A(P_i)$, $A(P_j)$ are the areas of the impact zones built along paths P_i and P_j. If $d(P_i, P_j) = 1$ there is a maximum dissimilarity, while $d(P_i, P_j) = 0$ means that there is a maximum similarity.

In order to solve a p-dispersion problem, one can use the Akgün et al. (2000) algorithm. The basic idea of this heuristic, divided in two phases, consists in building an initial solution (in a semi-greedy way) that is afterwards improved by means of a local search procedure. In the Akgün et al. paper, the description of this heuristic with interesting comments about its efficiency can be found.

The heuristic method is necessary since the p-dispersion problem is NP-complete and an exact algorithm for large-size problems would require a huge, sometimes unpractical, computational time.

The initial set U of candidate paths can be determined in several different ways; for example with the gateway shortest paths, the k shortest paths, or the iterative penalty method.

It is clear that the choice of the initial procedure will affect the quality (dissimilarity) of the obtained dissimilar paths, as well as the network topology and the origin–destination pair position will play a determining role in this sense.

For example, if one uses the k shortest-paths procedure followed by the p-dispersion method, one will obtain paths with dissimilarity indexes (both minimum and average) significatively greater than those associated with the paths generated by means of the Minimax method, even if this results in a greater average length.

Besides, the p-dispersion method wants to maximize the similarity (i.e., minimize the dissimilarity) while the Minimax method takes into account both dissimilarity and length of the generated paths. There is also another reason: the Minimax algorithm is a constructive greedy single-phase heuristic without a mechanism to escape from local optima, differently from the heuristic proposed by Akgün et al. that, instead, can rely on its second improvement phase.

Comparing the p-dispersion method based on an initial generation by means of the k shortest-paths procedure and on an IPM procedure, it can be observed that, in general, the former determines paths with a low dissimilarity but a with smaller length than the latter one.

The selection of paths that are too long can be prevented simply by forcing the IPM procedure to reject all paths whose lengths exceed a given threshold. Moreover, from a computational point of view, the IPM/p-dispersion method execution times are lower than the k-SHP/p-dispersion method ones.

The k-SHP/p-dispersion method in general behaves better than the IPM/p-dispersion method not only with respect to the length but also to the dissimilarity of the paths so found.

Summarizing, the experimental results quoted in Akgün et al. (2000) demonstrate that the dissimilarity objective may be reached easily if a p-dispersion method is applied to a large initial set of candidate paths. In general, however, this set has not to be necessarily generated with a procedure like IPM, that guarantees a high number of alternative initial dissimilar paths.

4.4.5 A Comparison Between a Multi-objective Approach and IPM

In this section, we will analyze the experimental results of a multi-objective approach and the IPM method, following what has been done in Dell'Olmo et al. (2005).

The multi-objective approach works in two phases: in the first phase np Pareto-optimal paths are generated (the minimization of both cost and risk is considered) between an assigned origin–destination pair; in the second phase, among the initial np paths, p spatially dissimilar paths, with $1 < p < np$, are selected with the p-dispersion method.

Phase 1: Minimization of cost and risk

Given an origin–destination pair, the set of Pareto-optimal routes (with respect to the cost and risk objective functions) is generated, using the bi-objective shortest-path algorithm proposed by Martins (see Chap. 2 for details on the algorithms); let np be the number of Pareto-optimal paths computed.

Phase 2: Equitable distribution of risk

1. The impact zones are designed along each path (the ray of an impact zone depends on the particular transported material).
2. The intersection areas among these impact zones are calculated.
3. With these data, a square dissimilarity matrix of size $np \times np$ is built; such a matrix is the input of a semi-greedy improvement heuristic that solves the p-dispersion problem (Akgün et al., 2000). This heuristic selects, among the np initial routes, p (with $1 < p < np$) dissimilar paths (according to the definition of dissimilarity adopted).
4. The desired number of dissimilar paths is then generated, assigning the requested value to parameter p.

Two weights (the objective function values) are associated with each arc: the length in km (as a measure of the transportation cost) and the risk index related to the average volume of traffic on the arc.

Martins' algorithm, even if easy to implement, could require huge computational effort both in the execution time and in memory-space consumption, due to the nature of the problem.

Tables 4.3 and 4.4 show on random graphs and grid graphs the CPU time needed to compute the efficient path set by means of a PC Pentium IV with 2.8 GHz and 512 MB of RAM. For random instances, the number of nodes n has been chosen equal to 100, 200, and 300, and the density d equal to 0.2, 0.5, and 0.7. For grid instances we considered a number of nodes equal to 10, and 20. Arc weights have been assigned uniformly at random in the range $[1,100]$. The initial and the terminal nodes of the instances are always 1 and n ($n \times n$ for the grid graphs), respectively.

Table 4.3 Results on random graphs (average over 10 instances)

		Total paths			Dominating paths			CPU time (s)		
n	d	min	ave	max	min	ave	max	min	ave	max
100	0.2	81	154.0	239	4	16.3	27	0.16	0.28	0.39
	0.5	547	875.6	1235	10	33.7	89	9.84	17.95	30.13
	0.7	1216	1587.2	2435	20	47.6	89	41.52	60.87	114.13
200	0.2	409	679.9	1022	14	35.7	79	23.2	52.07	120.18
	0.5	2008	3293.1	4255	24	56.0	74	754.08	1449.40	2281.29
	0.7	3960	5397.3	7004	34	68.0	110	2730.47	4620.08	7899.38
300	0.2	669	1406.6	2043	25	46.8	86	257.14	553.18	881.91
	0.5	4608	6331.5	9561	57	89.3	118	8349.63	12,442.50	25,560.15
	0.7	6956	10034.9	13469	57	109.9	160	19,221.06	32,210.43	55,705.37

Table 4.4 Results on grid graphs (average over 10 instances)

		Total paths			Dominating paths			CPU time (s)		
	n	min	ave	max	min	ave	max	min	ave	max
10×10	100	10	91.8	324	2	28.1	111	0.03	0.34	0.90
20×20	400	66	607.9	2667	16	167.5	832	41.03	1280.15	3617.34

In the following, we will show the results obtained considering a rectangular 11×16 grid-graph as a test problem with 176 nodes and 650 arcs. The risk and the length values were generated through a uniform distribution $(0,10)$ and $(0,1)$, respectively. The origin and the destination nodes considered are node 1 and node 176, respectively, that are the source and the sink in the grid. In this grid scenario, the calculation of the Pareto-optimal paths set has required only a few seconds and a moderate use of the memory space.

The dissimilarity index used is:

$$d(P_i,P_j) = 1 - \left\{ \left[\frac{L(P_i) \cap L(P_j)}{L(P_i)} \right] + \left[\frac{L(P_i) \cap L(P_j)}{L(P_j)} \right] \right\},$$

where $L(\cdot)$ is the length of the considered paths. This index says that two paths are more dissimilar if the length of the part in common is smaller.

The set of Pareto-optimal paths is formed by 41 paths; besides these paths, we also considered those with the highest and the lowest risk. The path with minimum risk has a value of 7.22, while that one with maximum risk has a value of 13.99.

The IPM method has been implemented assuming a multiplicative penalization with a factor of 1% (with respect to the last path found); the method generated 53 initial paths. The iterative penalty method has been stopped after having generated a path whose risk is greater than the path with maximum risk found by the multi-objective shortest path algorithm (value equal to 13.99).

The p-dispersion method was applied for $p = 3,\ldots,15$; the obtained results are shown in Tables 4.5 and 4.6 that show the minimum risk (*MR*) and average risk (*AvR*) values normalized with respect to the minimum risk value (i.e., 7.22), the minimum dissimilarity (*MR*) and the average dissimilarity (*AvR*). Note that these results appeared in the paper of Dell'Olmo et al. (2005).

By the analysis of the two tables it can be noted that with a multi-objective approach a lower risk level is distributed over quite similar paths, while with the IPM more dissimilar but more risky routes are obtained.

4.5 The Hazmat Transportation on Congested Networks

By the analysis of the literature presented in the first part of this chapter, the existence of a non-negligible work can be noted that focuses on the selection of routes between only one origin–destination pair and on the transportation of a single commodity at a time. These kinds of problems are identified as local planning problems. At a strategic planning level, it is necessary to consider the case in which different hazardous materials have to be shipped simultaneously among different origin–destination pairs. Such problems are identified as multi-commodity multi origin–destination transportation problems. These approaches seem to be the most promising ones because they allow a global planning of the transport activities in

Table 4.5 p-dispersion applied to the Pareto-optimal path set

p	MD	AvD	MR	AvR
3	0.615532	0.714331	1	1.377
4	0.567202	0.630178	1.11	1.386
5	0.505702	0.599593	1.008	1.415
6	0.372896	0.583096	1.044	1.391
7	0.352393	0.561606	1.115	1.410
8	0.335656	0.560406	1.115	1.437
9	0.301526	0.543386	1.008	1.396
10	0.282450	0.534498	1.027	1.384
11	0.252642	0.529762	1.044	1.369
12	0.237922	0.518066	1.044	1.337
13	0.186734	0.500093	1	1.309
14	0.181316	0.496556	1	1.321
15	0.15960	0.500615	1.027	1.308

Table 4.6 p-dispersion applied to the path set generated by IPM

p	MD	AvD	MR	AvR
3	0.987467	0.995288	1.65	1.646
4	0.968462	0.984351	1.32	1.625
5	0.928960	0.978973	1.329	1.600
6	0.905621	0.961741	1.457	1.654
7	0.849001	0.946629	1.329	1.565
8	0.824598	0.936090	1.330	1.597
9	0.774315	0.912843	1.259	1.563
10	0.770220	0.917116	1.259	1.553
11	0.768093	0.917362	1.259	1.571
12	0.724395	0.910029	1.259	1.582
13	0.716832	0.901485	1.044	1.551
14	0.663436	0.890317	1.008	1.550
15	0.662583	0.888165	1.044	1.562

terms of total costs, total risk and equity. In this context, in the following, two different multi-commodity approaches are presented for managing hazmat transportation with the aim of obtaining routes satisfying the equity criterion, that is guaranteeing spatially dissimilar paths. In particular, in order to guarantee a spatial distribution of the risk, the hypothesis of congestion will be considered, that is cost and risk will be also functions of all the number of vehicles traversing the transportation network.

4.5.1 Multi-commodity Minimum Cost Flow with and Without Congestion

As introduced previously, the minimization of cost and risk and the equity maximization are the main targets in hazmat transportation. In the following, we start by showing the standard formulation of the *multi-commodity minimum cost flow problem* in which the total cost function will be decomposed into two parts: one will consider the transportation costs, the other one can be defined as a risk cost depending on the transported material type. Such a cost can be interpreted as a penalty to pay (e.g, for deaths, injured people, interventions on the territory of the environment) in the case of an incident along a segment of the road network.

Although a standard multi-commodity flow model allows one to manage simultaneously different materials that have to be routed over the transportation network, the solutions of this model do not always guarantee fair solutions. In fact, as will be highlighted next, it can happen that the model fails in achieving the target of equity. This situation happens also because, in general, the model assumes that the network is unloaded, i.e., no other vehicle is traversing the network at the time of the assignment of a certain vehicle to the network arcs, and, therefore, the objective function value is independent on the presence of other vehicles transporting the same or a different material.

By considering the congestion assumption, a new multi-commodity flow model with a non-linear objective function will be defined that by means of additional variables and constraints can be eventually linearized. By a comparison between the multi-commodity flow model without and with congestion, it will be experimentally shown that this last formulation assures, for an origin–destination pair, a spatially dissimilar route set such that the equity goal can be achieved.

4.5.1.1 The Models Formulation

Multi-commodity flow models are used to plan the distribution activities of commodities on transportation networks. In particular, given an origin–destination pair and the amount of commodities to be transported between such an origin and a destination, the multi-commodity minimum cost flow model (MMCFM) finds the optimal distribution of such quantities minimizing the total transportation cost.

Let $G = (V, A)$ be the graph representing the transportation network; assume that the following quantities are assigned on each arc $(i, j) \in A$:

- p^k_{ij} the necessary cost to transport a unit of commodity k, with $k = 1, \dots, K$, along arc (i, j);
- x^k_{ij} the flow of commodity k, with $k = 1, \dots, K$, along arc (i, j);
- u^k_{ij} the maximum quantity of commodity k, with $k = 1, \dots, K$, that can be transported along arc (i, j);
- U_{ij} the maximum total flow of all commodities traversing arc (i, j); it represents the interactions among the different commodities that have to be transported.

Parameter b_k is associated with each vertex $v \in V$; it represents the supply or the demand of commodity k, with $k = 1, \dots, K$, and it allows one to define demand and supply nodes. In particular:

- $b^k_v > 0$ if v is a supply node for commodity k;
- $b^k_v = 0$ if in v the flow is conserved for commodity k;
- $b^k_v < 0$ if v is a demand nodes for commodity k.

MMCFM can be formulated as follows:

$$\min \sum_{(i,j)\in A} \sum_{k\in K} p^k_{ij} x^k_{ij}$$

$$\sum_{j:(j,i)\in A} x^k_{ji} - \sum_{j:(i,j)\in A} x^k_{ij} = b^k_i, \quad \forall i \in V, k = 1, \dots, K$$

$$x^k_{ij} \le u^k_{ij}, \quad \forall (i, j) \in A, k = 1, \dots, K$$

$$\sum_{k\in K} x^k_{ij} \le U_{ij}, \quad \forall (i, j) \in A$$

$$x^k_{ij} \in Z^+, \quad \forall (i, j) \in A, k = 1, \dots, K.$$

The first constraint is the balance constraint: which means that if i is an intermediate vertex ($b^k_i = 0$), then the flow emanating from i must be equal to the incoming flow in i; if i is the origin vertex ($b^k_i > 0$), then the net flow (the difference between the flow emanating from i and the flow incoming in i) must be equal to b^k; if i is a destination vertex ($b^k_i < 0$), then the net flow must be equal to $-b^k$.

The second constraint says that the flow x^k_{ij} of commodity k on the arc (i, j) must not to exceed the arc capacity associated to that commodity. The third constraint says that the flow of all the commodities traversing arc (i, j) must not exceed the arc total capacity.

In the hazmat transportation context, capacities u_{ij}^k (the maximum quantity of commodity k that can be transported along arc (i, j)) and U_{ij} (the maximum total flow of all commodities traversing arc (i, j)) can be interpreted as measures of the maximum amount of risk that one is willing to tolerate on arc (i, j), respectively referred to a single material k and to the total flow of materials traversing along arc (i, j).

Given the transport costs p_{ij}^k and the quantity b_i^k between assigned origin–destination pairs, MMCFM seems to be the most natural way to obtain a global flow distribution with minimum risk and cost.

In multi-commodity flow problems, the objective function minimizes only the transportation cost. When hazardous materials are concerned, it is better to decompose the cost p_{ij}^k into two main components: economic cost c_{ij}^k and risk cost t_{ij}^k, which can be seen as the cost to pay in the case of an incident. In particular, the latter represents the cost related to the consequences of an incident for the material k, and can be assumed to be proportional to the risk r_{ij}^k associated with arc (i, j).

So, the new total cost will be written as:

$$p_{ij}^k = c_{ij}^k + t_{ij}^k, k = 1, \ldots, K,$$

where the two items on the right-hand side are the economic cost and the risk cost, respectively, necessary to move a unit of commodity k through arc (i, j).

Another implicit assumption in MMCFM is that commodities are transported in an unloaded network, i.e., without taking into account the simultaneous presence of other vehicles in the network. This assumption, as we will see in the next section, limits the model to obtain a spatially dissimilar origin–destination path set, and, therefore, does not guarantee the equity target.

If we delete this assumption, we have to assume that the costs are functions of the flow traversing the arcs, that is:

$$p_{ij}^k(x_{ij}^k) = c_{ij}^k(x_{ij}^k) + t_{ij}^k(x_{ij}^k).$$

Put simply, we can assume that the economic cost is constant (so $c_{ij}^k(x_{ij}^k) = c_{ij}^k$), while a possible trend of function $t_{ij}^k(x_{ij}^k)$ may be the following:

$$t_{ij}^k(x_{ij}^k) = t_{ij,0}^k \left(1 + \frac{x_{ij}^k}{(u_{ij}^k - x_{ij}^k)}\right), \tag{4.1}$$

where $t^k_{ij,0}$, is the unitary risk cost for commodity k along the unloaded arc (i,j), while the ratio $\frac{x^k_{ij}}{(u^k_{ij}-x^k_{ij})}$ says that if the number of vehicles transporting a material k increases on the arc, then the risk and the respective cost will increase, becoming $+\infty$. So, $t^k_{ij}(x^k_{ij})$ can be seen as the cost related to the use of arc (i,j).

Even if by using function (4.1) one risks not considering the possible correlations and interactions among the different hazardous materials traversing the network, it is hard to foresee such interactions and to merge such relations in a model.

Now we have to define the role of parameters u^k_{ij} and U_{ij}. Referring to the model, x^k_{ij} has been generically defined as the flow of commodity k traversing arc (i,j). But if it represents the vehicle number transporting a particular hazmat k on arc (i,j), then u^k_{ij} and U_{ij} have to represent, respectively, the maximum number of vehicles transporting commodity k along the arc (i,j) and the total maximum number of vehicles flowing on arc (i,j).

Let $R_k = \max_{(i,j)\in A}\{r^k_{ij}\}$ be the maximum among risk values for commodity k on arcs. We can calculate for each arc $(i,j)\in A$

$$M^k_{ij} = \left[\frac{R^k}{r^k_{ij}}\right],$$

where $[\cdot]$ is the integer part operator, and we pose $u^k_{ij} = M^k_{ij}$.

This equation says that there is an inverse relation between the capacity u^k_{ij} and the maximum risk R^k_{ij}, so the vehicle number that can traverse the arc, transporting commodity k, has to gradually decrease as soon as the risk of a given arc approaches the maximum risk R^k_{ij}. Then, capacity U_{ij} can be expressed as a suitable linear combination among the capacities $u^k_{ij}, \forall k$.

Note that $t^k_{ij}(x^k_{ij}) = t^k_{ij,0}\left(1+\frac{x^k_{ij}}{(u^k_{ij}-x^k_{ij})}\right)$ is an increasing function of the hazardous materials traversing arc (i,j); hence, it is proportional to the risk level r^k_{ij} associated with arc (i,j) for each material k considered. Therefore, if the number of vehicles transporting hazardous material flowing on an arc increases, then the risk on that arc will increase, and the incident probability and its consequences will increase as well.

The formulation of a multi-commodity flow model with congestion will be:

$$\min \sum_{(i,j)\in A}\sum_{k\in K} c^k_{ij}x^k_{ij} + t^k_{ij,0}\left(1+\frac{x^k_{ij}}{(u^k_{ij}-x^k_{ij})}\right)$$

$$\sum_{j:(j,i)\in A} x_{ji}^k - \sum_{j:(i,j)\in A} x_{ij}^k = b_i^k, \ \forall i \in V, k = 1, \ldots, K$$
$$x_{ij}^k \le u_{ij}^k, \ \forall (i,j) \in A, k = 1, \ldots, K$$
$$\sum_{k \in K} x_{ij}^k \le U_{ij}, \ \forall (i,j) \in A$$
$$x_{ij}^k \in Z^+, \ \forall (i,j) \in A, k = 1, \ldots, K.$$

The target we propose with this model, and in particular with the use of the new objective function preventing the arcs saturation, is to provide solutions discouraging the use of a few arcs to transport a commodity k between its origin and destination nodes. That is, we would distribute the commodity k on more networks arcs. This implies to determine an equitable network flow that distributes risk (and so the incident consequence) among all the population.

4.5.2 Test Problems on Grid Graphs

The behavior of the two models (MMCF with and without congestion) has been evaluated on three different grid-like transportation networks of size 8×8, 10×10 and 12×12, respectively. We assumed to have only two materials to transport. For each arc (i, j) in the grid network, data related to risks and costs for each commodity have been generated at random using a uniform distribution Unif(0,1). The other parameters were generated based on the number of vehicles to be routed. For example, for the grids of size 8×8 and 10×10, we assumed a fleet of 5 and 8 vehicles, respectively, and generated capacities u_{ij}^k, with $k = 1, 2$, by means of a uniform distribution Unif(0,10). For capacity U_{ij}, a uniform distribution Unif(0,20) has been used for the grid 8×8, and a distribution Unif(0,100) has been used for the grid 10×10.

For the grid 12×12, we assumed a fleet of 10 vehicles and generated both the two capacity values u_{ij}^k and U_{ij} by means of a uniform distribution Unif(0,100).

For the grid 10×10 two origins have been considered, i.e., nodes 1 and 10, while we chose a unique common destination vertex, i.e., node 100. For the grid 12×12 the origins are nodes 80 and 84, while in this case we considered two destinations for each material, that are, respectively, nodes 1 and 133, and nodes 1 and 12.

In Table 4.7, we show the results obtained by both the two flow models. For each grid and each model, we give the percentages of the arcs used in the solution with respect to the total number of the arcs in the grid. This measure has been introduced because if the number of arcs in a solution is greater than another solution, then the number of paths onto which the materials are routed will be greater, and, hence, the path spatial dissimilarity will be greater too. The column Total refers to the total number of arcs used to transport both the two materials for each grid.

Table 4.7 Computational results on grids

Grid 8×8	Arcs used				
	Material 1	Material 2	Total	# of arcs used %	Cost
MMCF	26	25	51	22.75%	86.42
Congestion	37	32	69	30.80	117.35
Grid 10×10	Arcs used				
	Material 1	Material 2	Total	# of arcs used %	Cost
MMCF	29	9	38	10.56%	162,65
Congestion	71	41	112	31.11	229.71
Grid 12×12	Arcs used				
	Material 1	Material 2	Total	# of arcs used %	Cost
MMCF	21	36	57	10.80%	217.44
Congestion	29	76	105	19.87	230.78

The same conclusions observed in the previous section can be made for these test problems, that is the model with congestion is always able to obtain a spatially dissimilar paths set. In the particular case of grid 10×10, the multi-commodity flow model without congestion has determined only one path for one of the two considered materials. Such conclusions are also confirmed by the percentage of used arcs. Moreover, a trade-off between costs and equity is noticed. In fact, for the three grid-graph classes an increase exists of the transportation total cost of 5%, 41% and 6%, respectively, among the two models. Nevertheless, such an increase does not seem to be very significant and the solutions of the model with congestion satisfy the different needs of the actors (government and carriers) involved in the hazmat transportation problem.

4.5.3 The Linearized Model with Congestion

In this section, we present how to linearize the model with congestion. Such linearization, even if it has a number of variables and constraints greater than the non-linear model, allows one to use optimization software more efficiently than those developed for the non-linear programming problems resolution.

Given an arc (i,j) with capacity u_{ij}^k $k=1,\ldots,K$, let the variables

$$z_{ijm}^k, \; \forall m=1,\ldots,u_{ij}^k, k=1,\ldots,K, \forall (i,j)\in A$$

be such that

$$x_{ij}^k = \sum_{m=0}^{u_{ij}^k-1} z_{ijm}^k, \; \forall (i,j)\in A, k=1,\ldots,K.$$

The unit cost function can be rewritten as:

$$(c_{ij}^k + t_{ij}^k)x_{ij}^k + t_{ij}^k \left(\sum_{m=0}^{u_{ij}^k-1} \frac{(z_{ijm}^k m^2)}{(u_{ij}^k - m)} \right),$$

that is linear in the variables x and z. The new formulation of the problem will be:

$$\min \sum_{(i,j)\in A} \sum_{k\in K} (c_{ij}^k + t_{ij}^k)x_{ij}^k + t_{ij}^k \left(\sum_{m=0}^{u_{ij}^k-1} \frac{(z_{ijm}^k m^2)}{(u_{ij}^k - m)} \right)$$

$$\sum_{j:(j,i)\in A} x_{ji}^k - \sum_{j:(i,j)\in A} x_{ij}^k = b_i^k, \; \forall i\in V, k=1,\ldots,K$$

$$\sum_{k\in K} x_{ij}^k \le U_{ij}, \; \forall (i,j)\in A$$

$$x_{ij}^k = \sum_{m=0}^{u_{ij}^k-1} z_{ijm}^k, \; \forall (i,j)\in A, k=1,\ldots,K$$

$$\sum_{m=0}^{u_{ij}^k-1} z_{ijm}^k = 1, \; \forall (i,j)\in A, k=1,\ldots,K$$

$$0 \le x_{ij}^k \le u_{ij}^k, \; \forall (i,j)\in A, k=1,\ldots,K$$

$$z_{ijm}^k \in \{0,1\}, \; \forall (i,j)\in A, k=1,\ldots,K, m=0,\ldots,u_{ij}^k-1.$$

As we can note, for each arc (i,j) the number of introduced Boolean variables is $\sum_{k=1} + K u_{ij}^k$, so the total number of variables x and z is at most

$$|A| \max_{(i,j)\in A} \{\sum_{k=1}^{K} u_{ij}^{k}\} + K|A|,$$

where $|A|$ is the cardinality of the network arcs.

4.6 The Problem of Balancing the Risk

In this section, we discuss the problem of identifying the best way of balancing the risk for hazmat transportation on an available set of routes. Differently from what has been discussed in the previous section, in this scenario one does not seek spatially dissimilar routes because the path set is given and known *a priori*.

This problem is dealt with via a multi-objective approach aiming to minimize simultaneously both the transport cost and the risk, and distributing the hazmat flow in an equitable way over the population.

The models that will be presented belong to the class of multi-commodity flow models introduced in the previous section.

4.6.1 Problem Formulation

Let $G = (V,A)$ be a graph representing the transportation network, where V is the nodes set and A is the road arcs set. Let $a \in A$ be a generic arc of G. Assume that a capacity c_a is associated with each arc $a \in A$. Let OD be the set of origin–destination pairs with $OD = \{(o,d) : o,d \in V\}$, where each (o,d) pair is associated with a commodity with a positive demand $d_{od} > 0$.

Let P_{od} be the path set between origin o and destination d. $\hat{P} = \bigcup_{(o,d)\in OD} P_{od}$. Let

- $x_{od}(p)$ the flow on path $p \in P_{od}$;
- s_{od} the maximum amount of flow between o and d on paths P_{od}.

The goal is to minimize the maximum amount of flow s_{od} under capacity constraints, i.e.,

$$\min f(x,s) = \sum_{(o,d)\in OD} s_{od}$$

$$\sum_{p\in P_{od}} x_{od}(p) = d_{od}, \ \forall (o,d) \in OD$$

$$x_{od}(p) \leq s_{od} \quad \forall p \in P_{od}, \ \forall (o,d) \in OD$$
$$\sum_{(o,d) \in OD} \sum_{p \in P_{od}} \delta_{ap} x_{od}(p) \leq c_a, \ \forall a \in A$$
$$s_{od} \geq 0, x_{od}(p) \geq 0 \quad \forall p \in P_{od}, \ \forall (o,d) \in OD,$$

where $\delta_{ap} = 1$ is arc a belongs to path p for pair (o,d), and 0 otherwise.

The objective function and the second constraint can be rewritten as the following concave function:

$$\min \sum_{(o,d) \in OD} \max_{p \in P_{od}} \{x_{od}(p)\}.$$

When the arc capacities are specialized with respect to the different transported materials, i.e., by introducing c_a^{od} then the next constraint has to be added to the problem:

$$\sum_{p \in P_{od}} \delta_{ap} x_{od}(p) \leq c_a^{od}, \forall a \in A, \forall (o,d) \in OD.$$

A simpler formulation considers a unique upper bound s on the maximum flow quantity to distribute over all the paths for all materials as follows:

$$\min f^{max}(x,s) = s$$
$$\sum_{p \in P_{od}} x_{od}(p) = d_{od}, \ \forall (o,d) \in OD$$
$$x_{od}(p) \leq s, \ \forall p \in P_{od}, \forall (o,d) \in OD$$
$$\sum_{(o,d) \in OD} \sum_{p \in P_{od}} \delta_{ap} x_{od}(p) \leq= c_a, \ \forall a \in A$$
$$s \geq 0, x_{od}(p) \geq 0 \ \forall p \in P_{od}, \forall (o,d) \in OD.$$

Similarly to the previous case, the objective function with the second constraint of the problem can be rewritten as:

$$\min \max_{(o,d) \in OD} \max_{p \in P_{od}} \{x_{od}(p)\}.$$

We note that the previous formulations do not balance exactly the flows on the network; rather they minimize the maximum admissible flow quantity on all the paths between an origin–destination pair. It is possible to introduce slack variables $s_{od}(p)$, $\forall p \in P_{od}$, and $\forall (o,d) \in OD$. Hence, the constraints $x_{od}(p) \leq s_{od}$ and $x_{od}(p) \leq s$ in the above two models become, respectively:

$$x_{od}(p) - s_{od} + s_{od}(p) = 0; s_{od}(p) \geq 0, \forall p \in P_{od}, \forall (o,d) \in OD,$$

and

$$x_{od}(p) - s + s_{od}(p) = 0; s_{od}(p) \geq 0, \forall p \in P_{od}, \forall (o,d) \in OD.$$

So the two objective functions will be

$$\min f(x,s) = \sum_{(o,d)\in OD} \sum_{p\in P_{od}} \{s_{od}(p)\}.$$

From these formulations it is possible to determine different generalizations. Indeed, we can consider the case in which the set of paths is not known *a priori*; in this case, balancing must be achieved during the path construction. A way of dealing with this problem could be that of limiting the number of paths to be determined, choosing, for instance, only those that traverse highly risky zones. Therefore, in the previous formulations, the following constraints should be introduced:

$$\sum_{p\in P_{od}} y_{od}(p) \leq n_{od}, \forall (o,d) \in OD,$$

where

$$y_{od}(p) = \begin{cases} 1 \text{ if } x_{od}(p) > 0 \\ 0 \text{ otherwise,} \end{cases}$$

and n_{od} is the maximum number of paths to be determined. Moreover, it is possible to associate other costs with each arc a of the network. Let $w_a \geq 0, \forall a \in A$, be such costs, the corresponding path cost will then be given by

$$w_{od} = \sum_{a\in A} \delta_{ap} w_a.$$

4.7 Bi-level Optimization Approaches to Hazmat Transportation

Bi-level optimization approaches have been tackled in hazmat transportation typically to solve multi-objective optimization problems related to network design.

Kara and Verter (2004) considered a hazmat transportation network design problem for which they proposed a bi-level model involving both the carriers and the government authority objectives. The authors assumed that the carriers, represented by the follower decision maker, always use the cheapest routes on the hazmat transportation network designed by the government authority. The latter, which covers the role of the leader decision maker in the bi-level model, has the objective of selecting the minimum total risk network, taking into account the behavior of the carriers. In their model, hazardous materials are grouped into categories based on risk impact, and a network is designed for each group, without considering possible interactions among shipments of different categories. The bi-level integer programming problem is transformed into a single-level mixed-integer linear problem by replacing the follower problem by its Karush–Khun–Tucker conditions and by linearizing the complementary slackness constraints. Then, the mixed integer problem so obtained is solved by means of a commercial software. As also remarked in Erkut and Gzara (2008), the single-level mixed-integer linear model can fail in finding an optimal stable solution for the bi-level model (see the importance of the uniqueness of the solution of the follower problem explained in Chap. 2). In fact, in general, there are multiple minimum-cost routing solutions for the follower over the designed network established by the leader, which may induce different total risk values over the network. Kara and Verter (2004) did not discuss such an issue.

Erkut and Gzara (2008) considered a similar problem to that of Kara and Verter (2004), generalizing their model to the undirected network case and designing the same network for all the shipments. Erkut and Gzara considered the possible lack of stability of the solution of the bi-level model obtained by solving the single-level mixed-integer linear model, and proposed a heuristic approach that always finds a stable solution. Moreover, they extended the bi-level model studying the cost/risk trade-off by including the cost of the leader problem in the objective function.

Chapter 5
Central Business District Freight Logistic

Abstract In this chapter, we focus our attention on another problem arising in freight logistic and in particular in delivering goods to medium-sized stores in a central business district (CBD). One of the main concern in this kind of problems is that CBDs have limited off-street parking and loading/unloading operations are clearly strongly affected by this. This problem has been previously studied by Caramia et al. (2007) where it was pointed out how, in this scenario, freight distribution could be addressed from two perspectives: (i) from the viewpoint of delivery/pick-up firms, delivery itineraries need to be coordinated with consideration of the delivery capacities and times at store sites for parking, and loading/unloading operations; (ii) from the viewpoint of transportation and city planners, the "distribution capacity" in the CBD must be determined, including the average cost of distribution routes, the maximum number of routes that can be simultaneously coordinated, the total number of stores that can be served, etc., much in the way traffic engineers are interested in the "traffic capacity" of a transportation network under which the vehicles move efficiently. Both the above viewpoints were addressed in the paper of Caramia et al. (2007) by solving the problem of what delivery itineraries are available so that parking loading/unloading capacities and associated time windows are respected and the itineraries are "balanced" in a way that costs and number of deliveries fall in given ranges. In this chapter, we will present an extension of the latter work, paying particular attention to the multi-objective nature of the problem.

5.1 Introduction

It can be easily inferred how freight distribution in a congested urban network is influenced by many factors, like:

- the transportation network including traffic characteristics, parking capacity, delivery and pick-up facilities,
- type of clients being served,
- the type of goods being picked up or delivered.

In this chapter, we focus on the distribution problem involving medium-sized stores that is influenced not only by the location of the stores, but also by the size of the network links and the availability of parking areas for the freight vehicles. Indeed, where these medium-sized stores are located on narrow streets and with limited off-street parking, e.g., see central business district (CBD) it is hard to accommodate one or two trucks at a time for parking, loading and unloading operations, and even then the trucks may obstruct the street traffic. In fact, in such scenarios, some cities, for example London, Singapore and Milan, have started to impose congestion pricing in the CBD during peak hours so that there is less pollution and smoother traffic flows.

These location/transport characteristics have a strong influence on finding practical solutions to this problem. Each client requires different types of goods during the day, which usually are shipped by different trucks that cannot be unloaded simultaneously by them due to the shortage of parking areas. Yet, the preferred periods for the delivery are quite short and they are often during the same periods of the day. Therefore, these deliveries have to be scheduled according to multiple slots contained in just a few time windows for each customer.

The fleet involved to serve this type of demand is generally composed of vehicles that depart either from plants or distribution centers external to the CBD, possibly located in the hinterlands. Another important relevant factor is that often in these cities it is difficult to obtain logistical support for inter-modal distribution, where freight from large trucks is allocated to medium-size vehicles more suitable for the movement in the city center.

In such scenarios, from the viewpoint of the delivery/pick-up firms, one needs to develop delivery itineraries so that deliveries are coordinated with consideration of the delivery capacities and times at the customer sites for parking, loading/unloading operations; indeed, the available time windows for the delivery of service play a key role in finding the routes for the delivery/pick-up vehicles. On the other hand, from the viewpoint of transportation and city planners, there is an interest in determining the "distribution capacity" in the CBD, in much the same way the traffic engineers are interested in the "traffic capacity" of a transportation network under which the vehicles move efficiently. That is, the planners should be interested in the evaluation of the maximum freight distribution capacity of the CBD's facilities including the average cost of distribution routes, the maximum number of routes that can be simultaneously coordinated, the total number of stores that can be served, etc., with minimum disruption to other vehicular traffic; this could also assist them in setting congestion pricing schemes and other policies to enhance commerce.

Both the above viewpoints are addressed in this chapter by solving the following delivery-only problem: What delivery itineraries, from the distribution starting points (depots) to the customers, are available so that parking loading/unloading capacities and associated time windows are respected and the itineraries are "balanced" in a way that the number of deliveries fall in given ranges.

The paper of Caramia et al. (2007) studied the problem of analyzing the freight distribution capacity of an urban network and associated facilities providing a mathematical programming formulation, and to evaluate exactly the freight distribution capacity a branch-and-bound approach was proposed. Then, a heuristic working in an operational scenario and looking for high-quality delivery routes (in terms of average waiting time and costs) was developed.

In this chapter, we study the same problem in a multi-objective scenario. The outline is as follows. Section 5.2 contains a formal description of the problem and its mathematical programming formulation. In Sect. 5.3, we briefly recall the branch-and-bound algorithm proposed in Caramia et al. (2007), and describe the heuristic approaches. Computational results and comparison are then proposed. Conclusions are discussed in Section 5.4.

5.2 Problem Description

We are given a complete graph $G = (V,E)$, with node set V and edge set E, and a function $w : E \rightarrow R^+$ assigning a distance $w(e)$ with each $e \in E$. With each node $v \in V$ is assigned a multiple time window $tw(v,i) = [s(v,i), f(v,i)]$, $i = 1,\dots,n(v)$, where $s(v,i)$ and $f(v,i)$ are, respectively, the starting and the ending time of the ith time window of node v, and, $n(v)$ is the total number of time windows associated with v. The length of each time window is equal to the service time. A *service* $\gamma(v,i)$ is a couple $(v, tw(v,i))$; in the following, for the sake of brevity, we will denote a service just with γ.

A path π of length m is a sequence of m services, $\gamma_1, \ldots, \gamma_m$, such that

$$f(\gamma_k) + w(\gamma_k, \gamma_{k+1}) \leq s(\gamma_{k+1}), k = 1, \ldots, m-1, \tag{5.1}$$

where $s(\gamma_k)$ and $f(\gamma_k)$ are, respectively, the starting and the ending time of the time window associated with service γ_k, while $w(\gamma_k, \gamma_{k+1})$ is the weight of the arc connecting the two nodes associated with services γ_k and γ_{k+1}.

The duration of a path is the sum of

- the weights of its edges,
- the service times at each node,
- the waiting times (i.e., $s(\gamma_{k+1}) - (f(\gamma_k) + w(\gamma_k, \gamma_{k+1}))$). Two paths π_i and π_j are incompatible if they share the same service.

A path π is *feasible* if the following conditions hold:

a) its duration is not greater than a given threshold $T_{\max}$;
b) it does not visit more than one time window for each node;
c) it includes from $k_{\min}$ to $k_{\max}$ time windows.

In the following, unless otherwise stated, we refer to a set of paths as a set of compatible and feasible paths.

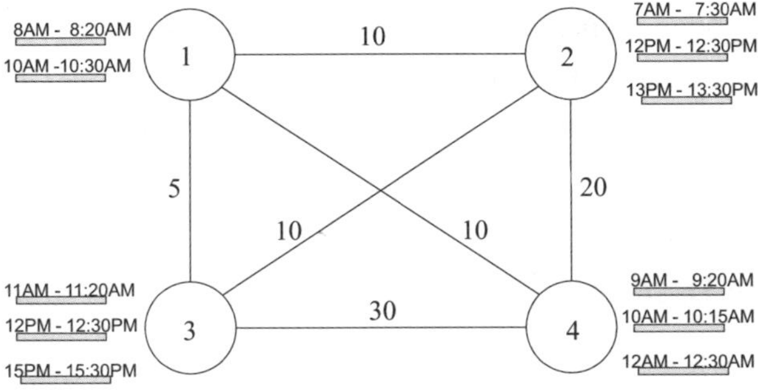

Fig. 5.1 A network with multiple time windows at nodes

Consider for example the simple network of four nodes in Fig. 5.1. The label on each arc is its weight. Node 1 has two time windows, while nodes 2, 3 and 4 have three time windows each. In total there are 11 services on the network. For example, service $\gamma(1,tw(1,1))$ is associated with the first time window $tw(1,1) =$ (8AM–8:20AM) of node 1. Let $T_{\max} = 400$ min, $k_{\min} = 2$, $k_{\max} = 4$ and consider the path $\pi_1 = \{\gamma(1,tw(1,1)), \gamma(4,tw(4,1)), \gamma(3,tw(3,1))\}$, that satisfies Eq. (5.1) and uses three services; it visits the first time window of node 1, the first time window of node 4, and the first time window of node 3. The waiting time at node 4 is equal to 30 min, and the waiting time at node 3 is equal to 1 h and 10 min. The duration of this path is given by the sum of the weights of its arcs, the service times and the waiting times at each nodes, that is: 10+30+20+20+20+30+70 = 200 min.

Path $\pi_2 = \{\gamma(2,tw(2,1)), \gamma(1,tw(1,1)), \gamma(2,tw(2,2))\}$ is not feasible since it does not respect condition (b). Path π_1 and path $\pi_3 = \{\gamma(2,tw(2,1)), \gamma(1,tw(1,1))\}$ are not compatible since they share service $\gamma(1,tw(1,1))$, while π_1 is compatible with path $\pi_4 = \{\gamma(4,tw(4,2)), \gamma(3,tw(3,2)), \gamma(2,tw(2,3))\}$.

Given a set of paths, we define the distribution capacity of the network as the total number of services visited by those paths. We look for a set of paths such that the distribution capacity is maximum.

5.2.1 Mathematical Formulation

In this section we provide a mathematical programming formulation for the problem. In order to do that, we define a new graph $G' = (V', E')$ as follows.

The new graph G'.

The node set of G' is $V' \cup \{\sigma_1, \ldots, \sigma_q\} \cup \{\phi_1, \ldots, \phi_q\}$, where q is an upper bound on the total number of feasible and compatible paths, each $i \in V'$ is associated with a service in G, and, $\sigma_1, \ldots, \sigma_q, \phi_1, \ldots, \phi_q$ are additional dummy nodes. Note that a trivial upper bound of parameter q is $\frac{\sum_{v \in V} n(v)}{k_{\min}}$. There is a (directed) arc (i, j) between two nodes $i, j \in V'$ if and only if the corresponding services satisfy constraint (5.1) and are associated with different nodes in G. The weight of any arc $(i, j) \in E'$, whose endpoints are the nodes associated with services i and j, is equal to the weight of the edge in G. E' also contains zero-weighted arcs between each σ_h, $h = 1,\ldots,q$, and each $i \in V'$, and zero-weighted arcs between each $i \in V'$ and ϕ_h, $h = 1,\ldots,q$.

A solution of the original problem described above is a set of paths $\pi_1,\ldots,\pi_q$, with origins in $\{\sigma_1,\ldots,\sigma_q\}$ and destinations in $\{\phi_1,\ldots,\phi_q\}$.

Consider the network in Fig. 5.2, with two nodes, one arc whose weight is equal to 40 and a total number of services equal to 6. Let $q = 1$. The corresponding new graph G' is depicted in Fig. 5.3. There are 8 nodes, 6 of them correspond to the services in the original network G and there are two additional dummy nodes σ_1 and ϕ_1. Dotted arcs in G' are the zero-weighted arcs; all the other arcs have weight equal to 40. Note that there is no arc among nodes 1 and 2, or among nodes 3, 4, 5, 6 since they correspond to services associated with the same node in G.

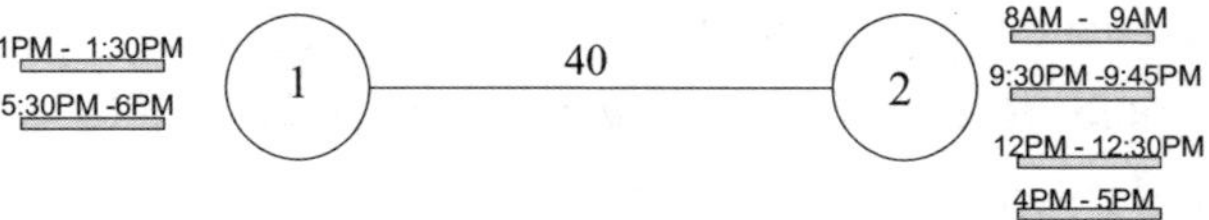

Fig. 5.2 The original network

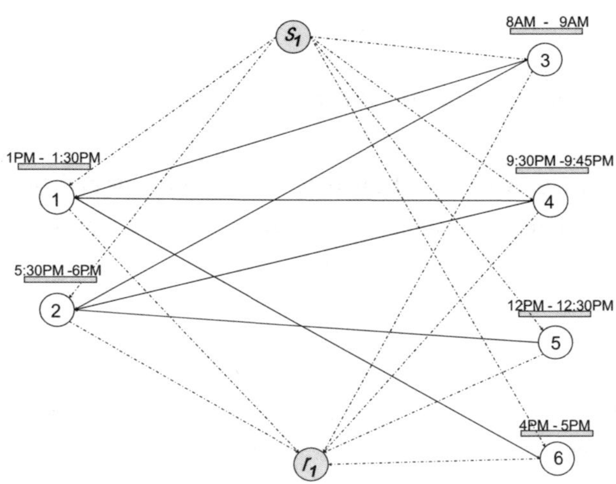

Fig. 5.3 The reduced network in the case of one path only

We are ready now to give the mathematical formulation of the problem. Let $\Sigma = \{\sigma_1,\ldots,\sigma_q\}$ and $\Phi = \{\phi_1,\ldots,\phi_q\}$. Moreover, let x_i^h be a binary variable equal to 1 if node i is in path h, and 0 otherwise. Similarly, let y_{ij}^h be a binary variable equal to

1 if arc (i,j) is in path h, and 0 otherwise. Furthermore, let $\Gamma(i)$ be the set of nodes reachable from i, and $\Gamma^{-1}(i)$ the set of nodes that reach i, $s(i)$ the starting time of the service associated with node $i \in V'$, and $f(i)$ its ending time, and, let $V_j \subseteq V'$ be the subset of nodes in V' corresponding to services of the original network G associated with the same node $j \in V$. The optimization problem (denoted in the following with $F1$) may be written as

$$\max z_1 = \sum_{i \in V'} \sum_{h=1}^{q} x_i^h \tag{5.2}$$

$$\text{s.t.} \quad x_i^h \leq \sum_{j \in \Gamma(i)} y_{ij}^h, \quad \forall i \in V', \quad h = 1, \ldots, q \tag{5.3}$$

$$\sum_{i \in V_j} x_i^h \leq 1, \quad h = 1, \ldots, q, \quad \forall j \in V \tag{5.4}$$

$$\sum_{i \in V'} x_i^h \leq 1, \quad h = 1, \ldots, q \tag{5.5}$$

$$\sum_{i \in \Gamma^{-1}(j)} y_{ij}^h = \sum_{w \in \Gamma(j)} y_{jw}^h, \quad j \in V', \quad h = 1, \ldots, q \tag{5.6}$$

$$\sum_{j \in \Gamma(i)} y_{ij}^h \leq 1, \quad \forall\, i \in \Sigma, \quad h = 1, \ldots, q \tag{5.7}$$

$$\sum_{i \in \Gamma^{-1}(j)} y_{ij}^h \leq 1, \quad \forall\, j \in \Phi, \quad h = 1, \ldots, q \tag{5.8}$$

$$k_{\min} \leq \sum_{i \in V'} x_i^h \leq k_{\max}, \quad h = 1, \ldots, q \tag{5.9}$$

$$x_j^h f(j) - x_i^h s(i) \leq T_{\max} \quad \forall i, j \in V', \quad h = 1, \ldots, q \tag{5.10}$$

$$x_i^h \in \{0,1\}, \quad \forall i \in V', \quad h = 1, \ldots, q \tag{5.11}$$

$$y_{ij}^h \in \{0,1\}, \quad \forall i, j \in V' \cup \Sigma \cup \Phi, \quad h = 1, \ldots, q. \tag{5.12}$$

The objective function z_1 maximizes the number of services covered by the paths. Constraints (5.3) assure that, given node i and path h, if there is no arc outgoing from i in that path then the corresponding service i is not visited (i.e., x_i^h must be set to 0); otherwise, since we are maximizing, if there is at least one arc outgoing from i, then x_i^h must be set to 1. Constraints (5.4) impose that in each subset of nodes V_j at most one service can belong to a path h. Constraints (5.5) say that a service can belong to at most one path. Constraints (5.6), (5.7) and (5.8) are classical relations to define a path. Constraints (5.9) allow paths to contain at least k_{min} services and at most k_{max} services. Constraints (5.10) limit the duration of each path to be at most T_{max}. Finally, we have integrality constraints on x_i^h and y_{ij}^h variables.

Consider now, the following multi-objective formulation (denoted in the following with $F2$), where the set of constraints is the same as in the previous formulation, except for the last one that allows one to evaluate the waiting time wt_{ij}^h at node j for service h if the previous node visited is i,

$$\min z_2 = \alpha \sum_{i \in V'} \sum_{h=1}^{q} (1 - x_i^h) + \beta \sum_{i \in V'} \sum_{h=1}^{q} wt_{ij}^h \tag{5.13}$$

$$\text{s.t. } x_i^h \leq \sum_{j \in \Gamma(i)} y_{ij}^h, \ \forall i \in V', \ h = 1, \dots, q \tag{5.14}$$

$$\sum_{i \in V_j} x_i^h \leq 1, \ h = 1, \dots, q, \ \forall j \in V \tag{5.15}$$

$$\sum_{i \in V'} x_i^h \leq 1, \ h = 1, \dots, q \tag{5.16}$$

$$\sum_{i \in \Gamma^{-1}(j)} y_{ij}^h = \sum_{w \in \Gamma(j)} y_{jw}^h, \ j \in V', \ h = 1, \dots, q \tag{5.17}$$

$$\sum_{j \in \Gamma(i)} y_{ij}^h \leq 1, \ \forall i \in \Sigma, \ h = 1, \dots, q \tag{5.18}$$

$$\sum_{i \in \Gamma^{-1}(j)} y_{ij}^h \leq 1, \ \forall j \in \Phi, \ h = 1, \dots, q \tag{5.19}$$

$$k_{\min} \leq \sum_{i \in V'} x_i^h \leq k_{\max}, \ h = 1, \dots, q \tag{5.20}$$

$$x_j^h f(j) - x_i^h s(i) \leq T_{\max} \ \forall i, j \in V', \ h = 1, \dots, q \tag{5.21}$$

$$x_j^h s(j) - (x_i^h f(i) + w[(i,j)]) = wt_{ij}^h, \ \forall i, j \in V', \ h = 1, \dots, q \tag{5.22}$$

$$x_i^h \in \{0,1\}, \ \forall i \in V', \ h = 1, \dots, q \tag{5.23}$$

$$y_{ij}^h \in \{0,1\}, \ \forall i, j \in V' \cup \Sigma \cup \Phi, \ h = 1, \dots, q. \tag{5.24}$$

Note that objective function z_2, differently from z_1, minimizes a linear combination of two objectives (α and β are two coefficients), one being the non-visited services, and the other is the total waiting time.

5.3 Solution Strategies

In this section, we report two solution approaches to $F2$. The former is the same as that presented in Caramia et al. (2007), and starts by solving the model $F1$ by means of a branch-and-bound algorithm, that is briefly recalled in the following.

The search tree of the branch-and-bound is formed by at most $|V' \cup \Sigma \cup \Phi|$ levels, and each level j of the search tree is formed by $j \cdot |V' \cup \Sigma \cup \Phi|$ nodes. Nodes in Σ and Φ are used, respectively, to open a new path, or to close one in the search tree. Using a depth-first search, one moves from a level to a subsequent level choosing the leftmost feasible service not yet chosen, i.e., a service that satisfies (5.1), and path-feasibility conditions a) and c) on $T_{\max}$ and $k_{\max}$. If at a certain level k there are ℓ paths in the current partial solution and the only choice is $\phi_{\ell+1}$ (that is, the current path $\pi_{\ell+1}$ has to be closed), the algorithm checks whether the current subpath $\pi_{\ell+1}$ has less than $k_{\min}$ services. If this is the case, the subpath $\pi_{\ell+1}$ cannot be considered in the current solution. Otherwise (i.e., the current subpath has at least $k_{\min}$ services) the path $\pi_{\ell+1}$ can be inserted in the current solution. In both cases, the algorithm goes on $\sigma_{\ell+2}$ on the next level $(k+1)$ to start a new subpath.

Once a subpath, say π_ℓ, is found, the algorithm checks whether it contains services of the same node. In this case, the algorithm retrieves feasibility by trying to split this subpath π_ℓ into two subpaths (note that the total number of visited services is the same). This is possible if and only if $2 \cdot k_{\min} \leq |\pi_\ell|$, where $|\pi_\ell|$ is the length of the current subpath π_ℓ. If this condition is not satisfied then the subpath π_ℓ cannot be considered in the current solution, and, as above, the algorithm goes on to the next level and starts a new subpath. The algorithm also uses some pruning rules.

The first rule orders arcs in E' starting with the cheapest one, so if one takes the first k arcs from the list we have a lower bound on the duration of a path of length k. Dynamically maintain this list so that when an arc is put in the current solution it is deleted from the list. Now, in the current subpath π_ℓ, let $r = k_{\min} - (|\pi_\ell| - 1)$, i.e., r is the difference between $k_{\min}$ and the number of arcs in the current subpath. If $r > 0$ do as follows: if the difference between $T_{\max}$ and the current subpath duration is lower than the sum of the r cheapest arcs in the list (which do not belong to any subpaths $\pi_1, \ldots, \pi_{\ell-1}$ in the current solution), then the subpath can be pruned.

Similarly, one can use a pruning rule at a node of the search tree as follows. Let t be the difference between the total number of nodes and the number of nodes considered in the current solutions (including those discarded by some of the above rules). If the sum of the current number of services plus t is smaller than the current upper bound value then the search tree can be pruned.

Note that with the root of the tree it is associated the linear relaxation of the integer program presented in the previous section. If the optimal solution of the linear relaxation is integer, then it represents an optimal solution to the integer variables; otherwise is offers an upper bound on the optimal solution value.

Once having solved problem $F1$ we have information on strategic parameters such as the capacity of the network, the average waiting time, and the average duration of a path.

Based on such results, the decision maker can improve the quality of the paths by means of a heuristic that tries to find the best paths in terms of minimum average total waiting time and minimum differences between path durations, by imposing, in a certain way, that paths must guarantee to have characteristics similar to those found by the exact algorithm. The heuristic is a greedy-type one and uses a depth-first visit of the graph. In particular, the algorithm selects the next node to be explored according to the minimum waiting time from the current node. The algorithm first builds a set of paths of length at least $k_{\min}$, then it performs a local search to improve the solution so found; and, after a certain number of iterations it removes some of the paths. Then, it restarts to build paths using the nodes exclusively belonging to removed paths. A detailed description of this heuristic, denoted with $H1$, is given in Table 5.1.

A second solution approach, denoted with $H2$, uses a slightly modify of the above heuristic, as described in Table 5.2.

5.3.1 Experimental Results

The codes have been implemented in the C language, while the linear relaxation has been solved with CPLEX version 10.

Results on the branch-and-bound algorithm are reported in Tables 5.3–5.5. It has been experimented on networks with 100, 500 and 1000 nodes, with $k_{\min} = 3$ and $k_{\max} = 3, 5, 8$, and $k_{\min} = 5$ and $k_{\max} = 5, 7, 8$. The service time has been defined as a Gaussian variable with mean value equal to 10 min. Each arc has a weight uniformly drawn at random from 1 to 20 min. Either 2 or 3 time windows have been

Table 5.1 Heuristic $H1$

Step 1. Select a node v from V' at random; find a path π starting from v, by adding each time a node with the smallest waiting time and that respects constraint (5.4), until either k_{max} is reached, or T_{max} is exceeded;
Step 2. If π has at least k_{min} services then remove the services of π from V' ;
Step 3. Repeat Step 1 and Step 2 until no more paths can be found;
Step 4. Let $desired\_length = \alpha * ave\_length$, with $0 < \alpha < 1$, be a percentage of the average path length, say ave_length, found by the exact algorithm;
Step 5. Let $desired\_wait = \beta * ave\_waiting$, with $0 < \beta < 1$, be a percentage of the average waiting time, say $ave\_waiting$, found by the exact algorithm;
Step 6. Try to improve the found paths as follows. For each path, try to replace a node with two available nodes, respecting constraints (5.4) and k_{max}, either if the total waiting time of that path is reduced or the total waiting time is less than $desired\_wait$;
Step 7. Repeat Step 6 until no more improvement is possible;
Step 8. Starting from the solution found, delete all the paths whose length is less than $desired\_length$, while holding all the others. Restart from Step 1, where V' is V' minus the set of nodes in the held paths, allowing the repetition of this restart until no more improvement is possible.

Table 5.2 Heuristic $H2$

Step 1. Select a node v from V' at random; find a path π starting from v, by adding each time a node selected by means of a Monte Carlo method that assigns to each candidate node a probability inversely proportional to its waiting time, until either k_{max} is reached, or T_{max} is exceeded (constraints (5.4) have to be obeyed);
Step 2. If π has at least k_{min} services then remove the services of π from V' ;
Step 3. Repeat Step 1 and Step 2 until no more paths can be found;
Step 4. Let $desired\_length = \alpha * ave\_length$, with $0 < \alpha < 1$, be a percentage of the average path length, say ave_length, found by the exact algorithm;
Step 5. Try to improve the found paths as follows. For each path, try to replace a node with two available nodes, respecting constraints (5.4) and k_{max}, either if the length of that path is augmented or the total length is greater than $desired\_length$;
Step 6. Repeat Step 5 until no more improvement is possible;
Step 7. Starting from the solution found, delete all the paths whose length is less than $desired\_length$, while holding all the others. Restart from Step 1, where V' is V' minus the set of nodes in the held paths, allowing the repetition of this restart until no more improvement is possible.

assigned to each node at random. Time windows have been generated from 7 A.M. to 11 A.M. Furthermore, T_{max} was set to 100 min.

In Table 5.3, we report, using average values on 10 test problems for each scenario, the percentage of covered services, the percentage of covered nodes, the average path duration, the average number of paths, the average number of services in a path, and the average waiting time of the paths.

As can be seen, the number of covered services (in per cent) is deeply effected by k_{min}. Indeed, this parameter determines the distribution capacity (i.e., the percentage of covered services) independently of the value of k_{max}. For example, on graphs with 100 nodes this percentage remains 98.2% when $k_{min} = 3$, and, 65.2% when $k_{min} = 5$. However, the number of paths found increases. That is, the same performance of the system can be achieved by means of different kinds of paths. This is because it is possible to cover remaining services at the low levels of the search tree by several numbers of short feasible paths.

On the other hand, k_{max} affects the quality of the paths. Observe the values of the average path duration, and the average waiting time in relation to k_{max}. With large values of k_{max} we have paths of longer average duration but shorter waiting time. In particular, waiting time is cut by a half in the instances with 1000 nodes when we move from instances with $k_{max} = 3$ to instances with $k_{max} = 8$.

Finally, for fixed values of k_{min} and k_{max}, the percentage of visited services of the network decreases as the size of the graph increases. This is due to the incompatibilities among time windows. Indeed, we generated time windows, as in real scenarios, to be concentrated in the same morning time intervals; therefore, the greater the size of the network, the more incompatibilities there are among services (i.e., the number of time windows that do not satisfy (5.1) increases). This behavior will be more evident looking at the results in Table 5.5, obtained by varying the average service time from 5 to 30 min.

Let us examine now Table 5.4 whose first four columns are the same as in Table 5.3. In column 5, we report the average optimal solutions of the linear relaxation of the formulation given in the previous section (which provide upper bounds on the optimal integer solutions) computed at the root of the branch-and-bound tree (again as average percentage values of covered services). The subsequent columns show the average number of times the pruning rules discussed in the previous section occur in the search tree, the average CPU time (in seconds), and the average number of branches in the search tree (Branches).

Table 5.3 Results of the exact algorithm with $T_{max} = 100$ min, the upper bound on the arc cost is equal to 20 min, and the average service time is equal to 10 min

Nodes	Number of services	k_{min}	k_{max}	Covered services (%)	Covered nodes (%)	Path duration (min)	Number of paths	Number of services in a path	Waiting time (min)
100	248.2	3	8	98.2	100.0	95.7	46.9	5.2	10.2
100	248.2	3	5	98.2	100.0	82.9	62.5	3.9	14.9
100	248.2	3	3	98.2	100.0	79.2	82.2	3.0	21.2
100	248.2	5	8	65.2	92.3	99.4	27.9	5.8	9.8
100	248.2	5	7	65.2	92.3	97.7	28.9	5.6	12.9
100	248.2	5	5	65.2	92.3	95.9	32.4	5.0	18.2
500	1278.3	3	8	92.2	100.0	94.3	226.6	5.2	12.5
500	1278.3	3	5	92.2	100.0	91.2	318.5	3.7	17.2
500	1278.3	3	3	92.2	100.0	87.1	392.9	3.0	25.9
500	1278.3	5	8	61.3	92.0	98.3	132.8	5.9	10.8
500	1278.3	5	7	61.3	92.0	96.5	142.5	5.5	13.7
500	1278.3	5	5	61.3	92.0	94.5	156.7	5.0	19.5
1000	2598.9	3	8	89.2	100.0	95.7	482.9	4.8	12.6
1000	2598.9	3	5	89.2	100.0	92.6	662.3	3.5	17.8
1000	2598.9	3	3	89.2	100.0	88.4	772.7	3.0	26.0
1000	2598.9	5	8	56.6	89.6	99.7	267.5	5.5	10.9
1000	2598.9	5	7	56.6	89.6	97.9	272.4	5.4	13.9
1000	2598.9	5	5	56.6	89.6	95.9	294.2	5.0	19.7

All statistics are averages of 10 test problems

Note that LP values follow the trend of the exact solution values, and exhibit the stability of the gap with respect to the different scenarios. The LP formulation is stable also from the computational point of view (observe the number of pivots in column 8 which grows linearly with the size of the problem instance).

As far as the behavior of the branch-and-bound algorithm is concerned, let us examine the last column of the table (Branches). It appears that more constrained instances require less branches (see, for example, those with $k_{min} = k_{max} = 3$). This is consistent also with the number of times prunings are executed (column 6) which follows the same criterion. Moreover, comparing the number of branches and the running times for instances with $k_{max} = 7$ and $k_{max} = 8$, we note that when $k_{max} = 7$ the number of branches is always larger than the latter case, while the running time is basically the same. In particular, we notice that for instances with 1000 nodes, the difference between the number of branches for $k_{max} = 7$ and $k_{max} = 8$ is greater than 5,000,000 while the CPU time is almost the same. However, this is not surprising and the reason for this behavior can be discovered by analyzing the number of times prunings are executed (column 6); this is always greater for $k_{max} = 7$ than for $k_{max} =$

Table 5.4 Performance of the exact algorithm with $T_{max} = 100$ min, the upper bound on the arc cost is equal to 20 min, and the average service time is equal to 10 min

Nodes	Number of services	k_{min}	k_{max}	Covered services (%) (LP relaxation at root)	Number of times a pruning is executed	CPU time (s)	Branches
100	248.2	3	8	99.6	1243.4	1121.2	10,423.2
100	248.2	3	5	99.6	1423.9	1102.3	15,253.1
100	248.2	3	3	99.6	802.1	983.2	7823.0
100	248.2	5	8	68.8	1429.9	1724.9	12,032.4
100	248.2	5	7	68.8	1912.0	1723.4	17,142.6
100	248.2	5	5	68.8	912.2	1598.2	10,243.7
500	1278.3	3	8	95.7	29,998.4	2512.8	333,536.1
500	1278.3	3	5	95.7	55,775.3	2510.1	488,096.4
500	1278.3	3	3	95.7	27,685.1	2429.5	250,336.2
500	1278.3	5	8	68.6	40,142.7	2772.6	385,024.7
500	1278.3	5	7	68.6	60,823.8	2789.2	548,544.9
500	1278.3	5	5	68.6	35,910.9	2699.1	327,776.3
1000	2598.9	3	8	93.3	989,002.2	7234.9	10,673,152.1
1000	2598.9	3	5	93.3	1,512,002.7	7129.0	15,619,072.2
1000	2598.9	3	3	93.3	901,203.7	7012.3	8,010,752.5
1000	2598.9	5	8	67.9	1,512,970.0	7457.6	12,320,768.0
1000	2598.9	5	7	67.9	2,078,223.5	7435.1	17,553,408.7
1000	2598.9	5	5	67.9	1,122,445.9	7199.4	10,488,832.2

All statistics are averages of 10 test problems

8. Indeed, the fact that the computation in both cases gave very similar CPU times, means that a single branch has a higher cost when $k_{max} = 7$ than when $k_{max} = 8$, as one would expect: that is, the greater the number of executed prunings, the lower the cost of each branch.

Table 5.5 shows the effect of the conflicts due to overlapping time windows on the distribution capacity of the system. With larger time windows the percentage of covered services decreases. Note that the gap between the minimum value (when average service time is equal to 5 min) and the maximum value (when average service time is equal to 30 min) also depends on the value of parameter $T_{max} = 100$ min.

The heuristics $H1$ and $H2$ have been tested on the same scenarios of the exact algorithm; the results are given in Tables 5.6–5.8.

A few seconds are required by the heuristics to find solutions on networks with 100 and 500 nodes, while on networks with 1000 nodes it runs in about 15 min. After several runs made by varying $\alpha, \beta \in \{0.75, 0.8, 0.85, 0.9, 0.95\}$, we set $\alpha = 0.85$ and $\beta = 0.85$ for which we obtained the best results that are shown in Table 5.6.

Table 5.5 Results of the exact algorithm with $T_{\max} = 100$ min., the upper bound on the arc cost is equal to 20 min, and the average service time varying in $\{5, 10, 15, 20, 25, 30\}$ min

Nodes	$k_{\min}$	$k_{\max}$	Service time (min)	Covered services (%)
100	3	6	5	99.5
100	3	6	10	98.2
100	3	6	15	92.5
100	3	6	20	90.9
100	3	6	25	85.9
100	3	6	30	40.7
100	5	8	5	80.4
100	5	8	10	65.2
100	5	8	15	50.5
100	5	8	20	45.1
100	5	8	25	15.2
100	5	8	30	6.8
500	3	6	5	94.8
500	3	6	10	93.5
500	3	6	15	88.1
500	3	6	20	86.6
500	3	6	25	81.8
500	3	6	30	38.8
500	5	8	5	76.6
500	5	8	10	62.1
500	5	8	15	48.1
500	5	8	20	43.0
500	5	8	25	14.5
500	5	8	30	6.5
1000	3	6	5	90.2
1000	3	6	10	89.1
1000	3	6	15	83.9
1000	3	6	20	82.4
1000	3	6	25	77.9
1000	3	6	30	36.9
1000	5	8	5	72.9
1000	5	8	10	59.1
1000	5	8	15	45.8
1000	5	8	20	40.9
1000	5	8	25	13.8
1000	5	8	30	6.2

All statistics are averages of 10 test problems

Table 5.7 compares the results of the heuristics with those of the exact approach. It shows the percentage differences between the values achieved by the exact and the heuristics for: the average number of visited services, the average number of paths

Table 5.6 Results of heuristic $H1$ with $T_{max} = 100$ min, the upper bound on the path duration is 20 min, and the average service time is 10 min

Nodes	Number of services	k_{min}	k_{max}	Covered services (%)	Covered nodes (%)	Path duration (min)	Number of paths	Number of services in a path	Waiting time (min)
100	248.2	3	8	95.4	100.0	96.2	34.8	6.8	5.2
100	248.2	3	5	94.8	100.0	90.2	51.2	4.6	6.9
100	248.2	3	3	92.4	99.8	85.4	76.4	3.0	15.2
100	248.2	5	8	63.5	92.2	99.5	24.2	6.5	5.8
100	248.2	5	7	62.8	92.1	98.7	24.7	6.3	7.9
100	248.2	5	5	61.2	91.9	97.9	30.4	5.0	14.2
500	1278.3	3	8	89.9	100.0	95.1	185.3	6.2	10.2
500	1278.3	3	5	87.2	100.0	93.4	237.2	4.7	12.1
500	1278.3	3	3	86.8	98.2	89.2	369.9	3.0	19.9
500	1278.3	5	8	58.1	91.3	98.7	109.2	6.8	8.2
500	1278.3	5	7	57.9	91.1	96.6	125.4	5.9	9.1
500	1278.3	5	5	57.3	87.5	94.8	146.5	5.0	14.5
1000	2598.9	3	8	85.4	100.0	96.2	389.4	5.7	9.8
1000	2598.9	3	5	83.7	100.0	93.5	494.4	4.4	11.8
1000	2598.9	3	3	83.4	98.1	91.1	722.5	3.0	17.2
1000	2598.9	5	8	52.9	89.6	99.8	211.5	6.5	8.9
1000	2598.9	5	7	52.1	89.2	98.2	229.5	5.9	9.7
1000	2598.9	5	5	51.5	89.2	96.9	267.7	5.0	11.2

All statistics are averages of 10 test problems

found, the average number of services in a path, the average path duration, and the average waiting time.

Refer now to Table 5.6 and Table 5.7. The trend of the percentage of the visited services decreases when the size of the network increases, as observed also for the exact algorithm in Table 5.3. Parameter k_{max} slightly influences the solution of the heuristic for a given network size and a given k_{min} value (on the contrary, recall that we observed in Table 5.3 that the percentage of visited services does not change when k_{max} changes for the exact algorithm).

The main differences between the heuristic and the exact algorithm are to do with the solution quality. Indeed, from the average waiting time (see column 10 in Table 5.6) it appears that heuristic solutions have a waiting time shorter that the exact solutions. The percentage decrement of the waiting time ranges from a minimum of 18% to a maximum of 54% (see column 8 in Table 5.7).

The decrement of the percentage of covered services is at most 9% (see column 4 in Table 5.7). This difference increases when the network size increases. For networks with 100 nodes the heuristic objective function value is at most 6% less than

Table 5.7 Comparison between $H1$ results and the exact approach results

Nodes	k_{min}	k_{max}	Covered services (decrement %)	Number of paths (decrement %)	Services in a path (increment %)	Path duration (increment %)	Waiting time (decrement %)
100	3	8	2.9	25.8	30.8	0.5	49.0
100	3	5	3.5	18.1	17.9	8.8	53.7
100	3	3	5.9	7.0	0.0	7.8	28.3
100	5	8	2.6	13.2	12.1	0.1	40.8
100	5	7	3.7	14.5	17.9	1.0	38.8
100	5	5	6.1	6.2	0.0	2.1	22.0
500	3	8	2.5	18.2	19.2	0.8	18.4
500	3	5	5.4	25.5	27.0	2.4	29.7
500	3	3	5.9	5.8	0.0	2.4	23.2
500	5	8	5.2	17.8	15.2	0.4	24.1
500	5	7	5.5	12.0	7.3	0.1	33.6
500	5	5	6.5	6.5	0.0	0.3	25.6
1000	3	8	4.3	19.3	19.1	0.5	22.2
1000	3	5	6.2	25.3	25.7	1.0	33.7
1000	3	3	6.5	6.5	0.0	3.1	33.8
1000	5	8	6.5	20.9	18.2	0.1	18.3
1000	5	7	8.0	15.7	9.3	0.3	30.2
1000	5	5	9.0	9.0	0.0	1.0	43.1

All statistics are averages of 10 test problems

the optimum, for networks with 500 nodes it is at most 7%, and, for 1000 nodes it is 9%.

The number of paths found by the heuristic is always lower than the number of paths found by the exact approach, and this decrement ranges from 5.8% to 25.8% (see column 5 in Table 5.7). This phenomenon is due to an increase of the average number of services in a path that (see column 6 in Table 5.7), when $k_{min} \neq k_{max}$, ranges from 7.3% to 30.8%.

Table 5.8 gives results on heuristic $H2$.

Note that, algorithm $H2$ finds solutions comparable with those of $H1$; in some cases $H2$ is also able to find paths with a larger number of covered services and nodes, as well as smaller waiting times.

5.4 Conclusions

We analyzed the problem concerning distribution of goods to medium-sized stores in a central business district. The objective of this problem is to plan and coordinate

Table 5.8 Results of heuristic $H2$ with $T_{max} = 100$ min, the upper bound on the path duration is 20 min, and the average service time is 10 min

Nodes	Number of services	k_{min}	k_{max}	Covered services (%)	Covered nodes (%)	Path duration (min)	Number of paths	Number of services in a path	Waiting time (min)
100	248.2	3	8	95.6	100.0	96.2	34.8	6.8	5.2
100	248.2	3	5	94.6	100.0	90.2	51.2	4.6	6.9
100	248.2	3	3	93.6	100.0	89.4	76.4	3.0	14.0
100	248.2	5	8	63.5	92.2	99.5	24.2	6.5	5.8
100	248.2	5	7	63.2	92.4	98.8	24.7	6.5	7.6
100	248.2	5	5	62.8	92.1	98.4	30.4	5.0	13.6
500	1278.3	3	8	89.9	100.0	95.1	185.3	6.2	10.2
500	1278.3	3	5	87.2	100.0	93.4	237.2	4.7	12.1
500	1278.3	3	3	87.2	99.4	91.4	360.6	3.0	18.4
500	1278.3	5	8	58.5	92.1	98.7	107.2	6.8	8.0
500	1278.3	5	7	58.2	91.8	97.0	124.2	5.9	8.6
500	1278.3	5	5	57.8	89.2	95.8	145.7	5.0	14.5
1000	2598.9	3	8	85.4	100.0	96.2	389.4	5.7	9.8
1000	2598.9	3	5	83.7	100.0	93.5	494.4	4.4	11.8
1000	2598.9	3	3	83.4	98.1	91.1	722.5	3.0	17.2
1000	2598.9	5	8	52.9	89.6	99.8	211.5	6.5	8.9
1000	2598.9	5	7	52.1	89.4	98.4	229.5	5.9	9.6
1000	2598.9	5	5	51.8	89.4	97.2	267.7	5.0	11.0

All statistics are averages of 10 test problems

distribution routes so as not to worsen the traffic congestion in the system. We analyzed the problem comparing the approach proposed in Caramia et al. (2007) with a slightly different one, evaluating both the approaches on several instances. As we stated in our previous paper, we remark that the analysis and the results presented can be applied by a central controlling or coordinating body that grants pick-up and delivery time windows. Thus, further research can be the analysis of both the amount of time and money saved for carriers planning routes according to our approach.

Chapter 6
Heterogeneous Staff Scheduling in Logistic Platforms

Abstract In this chapter, we analyze the problems arising in logistic parks management related to the operational resource scheduling problem motivated by the need to get the platform to operate according to the time and due dates of the distribution processes and balancing this objective with workers' shifts. In particular, we face a heterogeneous version of resource scheduling, where the term heterogeneous is referred to the existence of a pre-specified number of facilities qualifications that allow resources to be distinguished according to their capabilities (technical capabilities and operators skills). Roughly speaking, one is involved in assigning working shifts to a set of facilities and workers in a given planning horizon. The main difficulty lies clearly on the number of constraints that such shifts must obey. We devise an algorithm based on the paradigm of simulated annealing providing a large experimental analysis.

6.1 Introduction

In recent years the development of freight transportation infrastructures has been characterized by the establishment of a significant number of logistic platforms and very large logistic parks.

The amount of space of this single site set of facilities is impressive offering different services such us warehousing, packaging services, freight forwarders, customs clearance, transport and fuel. These have been motivated by the need to increase cargo carried on the rail network, to service the increasing road traffic across different national areas, and to improve access to ports and airports. Some projects focused on sites where shipments could be transferred from one mode to another, the so-called inter-modal nodes. At first it was as if the hinterland of every large city was growing a logistic platform, while nowadays very large logistic parks are seen as a service location for regional or national areas.

In Europe, one reason for this trend has been the change of barriers among the European countries and the increased number of European members. In this scenario, east–west distribution flows increased as well, and companies had to continually redesign and replan their distribution networks, change the cooperation with third logistic parties, and as a consequence rely more and more on the functions and facilities of logistic parks.

The relevant interest in the problem we are dealing with and the considerable amount of resources spent by several national and international governments are gearing the scientific community both toward the development of new models for the freight distribution on multi-modal networks, and toward the experimentation of these models on real-world case studies.

There have been several topics investigated in this research field, such as,

- routing of transportation vehicles (see, e.g., Barrett et al., 2006, and Boardman, 1999),
- the scheduling of containers (see, e.g., Hartmann, 2004, and Ewman and Yano, 2000),
- the assignment of traffic flows to a network (see, e.g., Aashtiani, 1979, Dafermos, 1982, and Bontekoning et al., 2004).

Bontekoning et al. (2004) analyzed more than 90 scientific publications concerning the various transportation problems existing for multi-modal networks and conclude by defining this topic as a completely new research field. Among the open problems, Macharis and Bontekoning (2004) and Jang and Kim (2005) point out that the problem of inter-modal freight routing has not been well studied yet in the optimization field and, more generally, there is still a widespread lack of optimization techniques to be used in a multi-agent context.

Even if there is a general acknowledgement of the scientific literature toward the advantages of using a multi-agent approach for modelling multi-modal systems (Hansen, 1979, and Davidsson et al., 2004), the major fraction of scientific publications on this topic concern problems related to the architecture of these systems, with a strong emphasis on the negotiation and communication protocols used by different agents (Ehrgott and Gandibleux, 2000, and Ehrgott and Gandibleux, 2002).

The currently developed and implemented optimization approaches mostly deal with mono-modal-only transportation networks, involving one kind of transportation vehicle per time (Alicke, 2002, and Balobo and Pinson, 2005). Instead, there is still much room for improvement on topics such as the allocation and the usage of the resources of a multi-modal transportation network (Kefi et al., 2007, and Zhu, 1999).

The scientific contributions cited so far are mostly theoretical, whereas there is a general lack of software environments for supporting the analysis of these problems from the experimental point of view and on data sets coming from the real world. One attempt, still underway, of creating an integrated system for the development and the experimentation of models and algorithms for the multi-modal freight transportation is the one made by Crainic et al. (2007). Their proposal, at the moment still in a conceptual phase, consists of an environment in which the different multi-modal transportation strategies will be implementable and analyzable according to several different aggregation dimensions. One point they stress to this end, is the need for evaluating the different transportation strategies in a time interval long enough to make it possible for congestion phenomena to arise in the transportation network.

A better situation exists instead in the field of the software that can be used for building multi-agent systems. In this case, there are several documented experiences about multi-agent systems that have been actually implemented and experienced for solving logistic transportation problems, such as Bergkvist et al. (2005), Buchheit et al. (1992), and Bürckert et al. (2000). However, as pointed out in Davidsson et al. (2004), most of these experiences refer to small-scale experimentation that have been conducted without using data and problems coming from the real world.

Usually, one way to see the efficiency of a complex structure like a large logistic park is measuring the throughput in terms of volumes of goods. Of course, that kind of measure depends directly on the time the facilities are able to operate in time, say on a week horizon. Because of the huge investments and the necessity to respect due dates of deliveries one would try to minimize the time the resources are idle. As most of the facilities need, in general, a group of workers with different skills to operate, the problem of efficiency of the equipments is intimately connected with the capability of assigning shifts to workers, balancing efficiency with workers shifts constraints.

The general problem of allocating shifts to workers is known as the staff-scheduling problem. In this scenario we study a particular kind of staff-scheduling problem known as the *constrained heterogeneous workforce scheduling problem* (CHWSP) (Tanomaru, 1995). The problem is defined as follows. Given a planning horizon T, a set E of workers classified into groups $q \in Q$ according to their qualifications, a number of constrains on the length of the working shifts and on how these shifts should be spaced out in the planning horizon, one is involved in finding a working schedule for each employee obeying the mentioned specifics.

Staff scheduling is very difficult to solve exactly since it is known to be NP-complete (Bartholdi, 1981). Consequently, approaches based on the design of efficient heuristics could be worthwhile, especially considering the number of constraints involved in the problem and the large-sized instances deriving from real applications.

Different versions of staff-scheduling problems have been examined in the scheduling literature. Trivedi (1981) studied a nurse scheduling problem that includes multiple objectives and full-time, part-time, and over time nurses. Loucks and Jacobs (1991) studied the staff-scheduling problem with workers who differ in their times of availability and task qualifications. For this problem, the authors proposed a heuristic procedure capable of producing a high-quality solution for a real-life problem in the presence of two objectives. The first one is related to the manager's concern and is the minimization of overstaffing; the second objective reflects the workers concern and is the minimization of the sum of the (squared) differences between the number of work hours scheduled and the number targeted for each employee. Constraints including minimum and maximum shift lengths and a maximum number of workdays are considered in the model.

Narasimhan (1997) faced a single shift-scheduling problem, which contains a heterogeneous workforce whose capabilities can be structured hierarchically. In Brusco and Jacobs (1993), Easton and Mansour (1993), Jarrah et al. (1994), and Tanomaru (1995) the authors apply metaheuristic algorithms, i.e., genetic operators, simulated annealing and tabu search, to solve various versions of staff-scheduling problems.

Cai and Li (2000) studied a multi-objective version of the staff-scheduling problem with mixed skills. The first objective considered is the minimization of the total cost for assigning staff to satisfy the manpower demands over time; the second objective is to find a solution that maximizes the surplus of staff among the solutions with almost the same value of the first objective; the third objective is to reduce the variation of staff surplus over different scheduling periods. For this problem the authors proposed a genetic algorithm approach.

In this chapter, we tailor an algorithm for the CHWSP based on the paradigm of simulated annealing. Refer to Chap. 2 of this book for simulated annealing in a multi-objective scenario. The outline is as follows. Section 6.2 describes the characteristics of the studied problem in terms of constraints and objective functions. Section 6.3 introduces the simulated annealing technique. In Section 6.4, we describe how we have engineered the simulated annealing algorithm for staff scheduling, and, in Section 6.5, we present and analyze experimental results.

In general, the workforce-scheduling problem is one among the complex set of optimization tasks associated with a logistic platform planning problem. As stated in the paper of Carrera et al. (2008), we can distinguish three families of tasks and related constraints: negotiation, that occurs at a planning level, storage, that occurs at a planning and a scheduling level, and human workforce, that occurs in a different way at both levels too.

The flow of products that arrives in a platform from a supplier can be divided into two categories: one formed by the products that remain in the platform and therefore requires storage, and the other formed by the products that have to be delivered immediately to the clients. In general, the former products represent a small amount of the total, since they mainly play the role of replenishing the products that have to be delivered to clients.

At the human workforce level we have to consider both the transportation activities that link the platform with the vending centers and the production activities related to carry out the operations in the platform, e.g., storage activities. In the following, we will concentrate on this second aspect, since the former can be modelled as a vehicle-routing problem, that is present in the literature in many different variants (see also the introduction chapter of this book).

Workforce scheduling in a logistic platform is extremely useful because of the extra workers and the overtime costs that could be generated when peaks of human workforce occur in some period of the year. To give an example, in Carrera et al. (2008) the authors considered an application of a platform that receives product from its suppliers twice a year for about two million articles crossing the platform to serve 90 clients. During these two periods overtime and additional temporary workers are needed and the authors studied the problem of how to smooth the level of workforce to minimize costs related to the earliness and tardiness penalties for product delivery and for extra workers and overtime costs.

6.2 The Heterogeneous Workforce Scheduling Problem

A working schedule can be represented as a list of working shifts, where each shift is defined by its starting time and the finishing time. For an employee $e \in E$ the ith working shift (denoted with s_i^e) is defined as

$$s_i^e = (t_{i1}^e, t_{i2}^e),$$

where t_{i1}^e and t_{i2}^e denote the starting time and the finishing time of the ith shift of the employee e, respectively.

The schedule of employee e is therefore a linked list of working shifts. In particular, assuming that ℓ is the number of working shifts for employee e during the planning horizon, the global working schedule for e is

$$s^e = \left[s_1^e, s_2^e, \ldots, s_{\ell-1}^e, s_\ell^e\right].$$

Consequently, the working schedule of all the employees can be denoted as

$$s = \left[s^1, s^2, \ldots, s^{|E|-1}, s^{|E|}\right].$$

Consider one week as a planning horizon and the hour as a time unit; the working shift planning must be carried out in a time horizon constituted by 168 discrete, consecutive and uniform time intervals. Fig. 6.1 reports an example of a solution for the case in which $\ell = 8$.

Throughout the chapter we will assume, without loss of generality, that the hour is the time unit and the planning horizon is one week.

6.3 Constraints and Objective Functions of the CHWSP

Here follows the description of all the constraints considered. Successively, we will discuss about the objective functions.

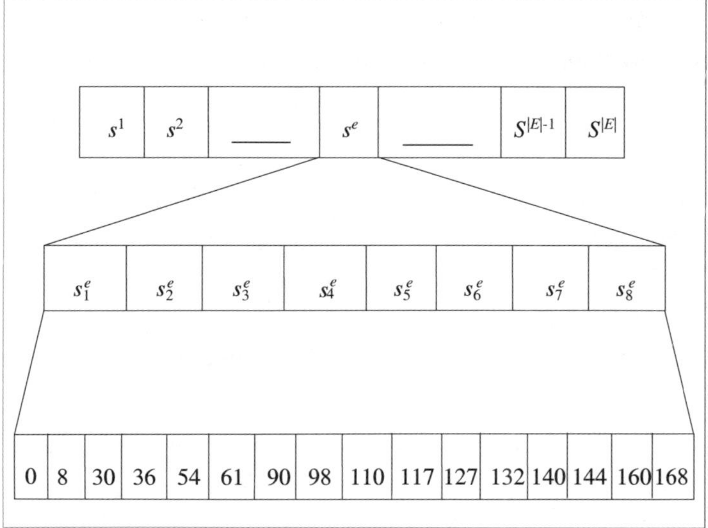

Fig. 6.1 Example of a solution of the CHWSP with $|E|$ employees and maximum number of working shifts $\ell = 8$

6.3.1 Constraints

The constraints considered for the problem can be divided into two categories:

- *Intra-schedule constraints*, i.e., constraints that concern the existing interactions into the working schedule of each employee;
- *Extra-schedule constraints*, i.e., constraints that concern the existing interactions among the working schedules of all the employees.

Referring to *intra-schedule* constraints we have:

1. Constraints associated with the overlap of the shifts, i.e., the finishing time of a shift must be earlier than the starting time of the successive shift. Analytically, such constraints can be defined as follows

$$t_{i,2}^e \leq t_{i+1,1}^e.$$

   These are the only hard constraints considered in the problem.
2. Constraints on the minimum duration of a working shift. Let $L_{\min}$ be the minimum duration of a working shift, then

$$t_{i,2}^e - t_{i,1}^e \geq L_{\min}.$$

3. Constraints on the maximum duration of a working shift. If $L_{\max}$ is the maximum duration allowed for each working shift, we have

$$t^e_{i,2} - t^e_{i,1} \leq L_{\max}.$$

4. Constraints on the minimum number of working hours in the whole planning time horizon, i.e.,

$$\sum_{i=1}^{\ell} (t^e_{i,2} - t^e_{i,1}) \geq h_{\min},$$

where $h_{\min}$ is the minimum number of working hours for each employee during a week.
5. Constraints on the maximum number of working hours in the whole planning horizon, i.e.,

$$\sum_{i=1}^{\ell} (t^e_{i,2} - t^e_{i,1}) \leq h_{\max},$$

where $h_{\max}$ is the maximum number of working hours for each employee during a week.
6. Constraints on the minimum idle time between two consecutive shifts ($R_{\min}$), i.e.,

$$t^e_{i+1,1} - t^e_{i,2} \geq R_{\min}.$$

Referring to *extra-schedule* constraints, we considered the minimum number of active employees each time of the week. Let $A_{\min}$ be such a value; we get the following relation

$$\alpha_t \geq A_{\min},$$

where α_t is the number of working people at t.

6.3.2 Objective Functions

The objective functions are different cost functions. Such functions can be grouped into two categories:

- Basic costs;
- Costs associated with the violation of the constraints introduced above.

Basic costs

If we denote with w_e the per hour cost of each employee, then the basic cost of the schedule of a generic worker e can be expressed as

$$c_{bi}^{e} = \begin{cases} d_i^e w_e \text{ if } d_i^e \leq L_{\max} \\ L_{\max} \text{ else,} \end{cases}$$

where $d_i^e = t_{i,2}^e - t_{i,1}^e$ is the ith shift duration associated with the employee e. Consequently, the basic cost is

$$c_b^e = \sum_{i=1}^{s} c_{bi}^{e} w_e.$$

Then, the basic cost of all the employees is

$$C_b = \sum_{e=1}^{|E|} c_b^e.$$

Constraint violation costs

A CHWSP solution can violate some constraints; for instance, it is allowed that some employees can work for a greater number of shifts with respect to the maximum established ones (overtime hours). Such violations are penalized by increasing the scheduling cost according to the degree of violation. In the following, we show the cost types appearing in the objective function.

1. The cost v_1^e associated with the violation of the established minimum number of working hours for each working shift $h_{\min}$. The working shift cost is increased by a quantity proportional to the difference between $h_{\min}$ and the number of real working hours. Indicating the weight of the violation with w_1, we can write

$$v_1^e = w_1 w_e \sum_{i=1}^{\ell} \left[L_{\min} - (t_{i,2}^e - t_{i,1}^e) \right].$$

2. The cost v_2^e associated with the violation of the maximum number of working hours for each working shift h_{max}. Such a cost is increased by the difference between the number of scheduled hours of the shift and $h_{\max}$, weighted with a coefficient w_2, i.e.,

$$v_2^e = w_2 w_e \sum_{i=1}^{\ell} \left[(t_{i,2}^e - t_{i,1}^e) - L_{\max} \right].$$

3. The cost v_3^e associated with the violation of the minimum number of weekly working hours. Such a cost is increased by a quantity proportional, by means of a weight w_3, to the difference between the minimum number of weekly hours and the number of weekly hours scheduled for each employee

$$v_3^e = w_3 w_e \left[h_{\min} - \sum_{i=1}^{\ell} (t_{i,2}^e - t_{i,1}^e) \right].$$

4. The cost v_4^e of the violation of the maximum number of working hours during the planning horizon. Such a cost is increased by a difference between the number of scheduled working hours in the week and the maximum number of established hours, weighted with w_4, i.e.,

$$v_4^e = w_4 w_e \left[\sum_{i=1}^{\ell} (t_{i,2}^e - t_{i,1}^e) - L_{\max} \right].$$

5. The cost v_5^e of the violation of the minimum number of idle periods among two consecutive shifts. Such a violation implies an incremental cost equal to the difference between the minimum number of idle periods and the number of idle times between two consecutive shifts. Thus,

$$v_5^e = w_5 w_e \left[R_{\min} - \sum_{i=1}^{\ell} (t_{i+1,1}^e - t_{i,2}^e) \right],$$

where w_5 is a proper violation weight.

6. The cost v_6^e associated with the violation of the minimum number of active employees at each time t in the planning horizon,

$$\begin{cases} w_6 \sum_{t=0}^{168} [A_{\min} - \alpha_t] \text{ if } \alpha_t < \gamma R_t \\ \sum_{t=0}^{168} [A_{\min} - \alpha_t] \quad \text{otherwise,} \end{cases}$$

where the coefficient $\gamma \in [0, 1]$.

Scalarized objective function

The cost of a feasible solution is given by the sum of the introduced costs. If we consider the working schedule of the employee e its cost is

$$c^e = c_b^e + \sum_{k=1}^{6} v_k^e.$$

Then, the cost of a global working schedule is

$$C = \sum_{e=1}^{|E|} c^e.$$

Therefore, the CHWSP scalarized objective function is

$$\min C = \sum_{e=1}^{|E|} c^e.$$

6.4 Simulated Annealing: General Presentation

Simulated annealing (SA) techniques arise from the existing analogy between the physical process of annealing and the mathematical process of searching for solutions in combinatorial optimization problems. In particular, this mathematical process is comparable with the cooling process of a solid. Moreover, an analogy exists among the costs of the objective function of the optimization problem, which converges to a minimum value (be it local or global), and the convergent states of the solid towards an equilibrium state, characterized by a minimum energy. In fact, the principal parameter that characterizes the annealing technique and determines its convergence is called *temperature*, or *control parameter*. At a high temperature, the atoms are free to vibrate around their rest positions and their movement is chaotic. At sufficiently low temperatures, vibration is no longer possible and the configuration of the atoms is called *frozen*. If the solid is cooled quickly, the possibility to get a stable structure is very small. Indeed, the cooling process has to evolve in a very slow way to allow the material to pass through continuous conditions of equilibrium. Therefore, the cooling process of a solid can be described as follows: for every value of the temperature T, the solid must achieve thermal equilibrium. The probability that the solid is in a state s with energy $E(s)$ is described by the

Boltzmann distribution:

$$P_T(s) = \frac{\exp\left(\frac{-E(s)}{kT}\right)}{\sum_w \left(\frac{-E(w)}{kT}\right)}$$

where k is the Boltzmann constant and w is a possible state. For large values of the temperature all energy states are possible, but as the temperature decreases, for the Boltzmann distribution, only the states with smaller energy are possible. When the temperature is close to 0, only the states with minimum energy have a probability different from zero. The idea to associate such a physical process with an optimization procedure arises from the consideration that the latter can be viewed as a number of *objects* that have to be scheduled to minimize an objective function. The vibrating atoms are replaced by such objects, and the objective function replaces the energy of the system, as illustrated in Table 6.1.

Table 6.1 Parallelism between the parameters of a combinatorial optimization problem and of a thermodynamic simulation problem

Thermodynamic simulation	*Combinatorial optimization*
States of the system	Admissible solutions
Energy	Cost
Change of the state	Neighbor solution
Temperature	Control parameter
"Frozen" state	Heuristic solution

Referring to Table 6.1 and to the above description, if we indicate with $C_i(k)$ the objective function cost at iteration k, the SA algorithm does not search for the best solution in the neighborhood of the current solution, but chooses *at random* a solution x in the neighborhood of the current solution i. Therefore, if $C_j(k)$ is less than $C_i(k)$, then j becomes the new current solution. Otherwise, j becomes the new solution with probability $P_{ij}(k)$, given by the Boltzmann distribution (Pirlot, 1996)

$$P_{ij}(k) = \frac{\exp(-\Delta C_{ij}(k))}{c_k},$$

where c_k is the control parameter value at iteration k and $\Delta C_{ij}(k) = C_j(k) - C_i(k)$. Typically, the probability $P_{ij}(k)$ decreases with k and with the size of $\Delta C_{ij}(k)$. As can be inferred from the description of the SA in the previous subsection, the latter is essentially governed by the following four parameters

1. The length of the Markov chains;
2. The first value of the control parameter, c_0;
3. The final value of the control parameter, c_f. Such a value is determined through the so-called *stop criterion*;
4. The rule for changing the value of the control parameter from c_k to c_{k+1}.

Some formalism for determining the above-mentioned parameters will now be described. In particular, next we will introduce Markov chains, whose length is used to determine the number of iterations that can be performed with the same value of control parameter; then, we will describe how to determine the initial value and the final value of the control parameter, and, finally, some decrement rules for the parameter control will be mentioned.

6.4.1 The Length of the Markov Chains

The mathematical model for the SA algorithm analysis is based on Markov chains. In fact, the changing process of a current configuration into the neighbor configuration is well described by a Markov chain. Such a chain is a succession of transitions in which the result of every trial is independent of the precedent states. A Markov chain is defined by a transition matrix of the states and by an initial distribution. Therefore, given a finite set of states $N = \{1,\dots,n\}$, where each state is a possible configuration, the probability P_{ij} to pass from a state i to a state j satisfies the following conditions

$$P_{ij} \geq 0 \quad \forall (i,j) \in N \times N$$

$$\sum_{j \in N} P_{ij} = 1 \quad \forall i \in N.$$

In addition, the matrix of transition is

$$\begin{pmatrix} P_{11} & P_{12} & \dots & P_{1n} \\ P_{21} & P_{22} & \dots & P_{2n} \\ \dots & \dots & \dots & \dots \\ P_{n1} & P_{n2} & \dots & P_{nn} \end{pmatrix}.$$

In a SA algorithm the transition probabilities at an iteration k must depend on the value of the control parameter c_k. When T is a constant, a homogeneous Markov chain is obtained and the transition matrix is defined as the product of the probability to generate a configuration j from a configuration i, and the probability to accept the configuration j. SA is an algorithm that can be described by a sequence of an homogeneous Markov chain. Such a chain is generated by a constant value of the control parameter that is decreased through subsequent chains. If a SA algorithm is described by only one inhomogeneous Markov chain, in which the value T of the control parameter is decreased through subsequent transitions, such an algorithm is called an inhomogeneous SA. The simplest choice to determine the length of the Markov chains is to fix for those a value that depends on the problem size, as illustrated in Bonomi and Lutton (1984), Bonomi and Lutton (1986), and Lutton and Bonomi (1986). Therefore, the length L_k is independent of k. A more elaborated way of determining the length of Markov chains is based on the minimum number of transitions that should be approved for each value of the control parameter c_k. Such a method introduces some limits. In fact, when c_k tends to zero, the number of approved transitions decreases and the length of the Markov chains would tend to infinity. Consequently, the choice of a constant value for L_k is preferred, at least for low values of c_k.

6.4.2 The Initial Value of the Control Parameter

The initial value c_0 of the control parameter is determined so virtually all the transitions are approved. This means that c_0 is calculated so $\exp\left(\frac{-\Delta C_{ij}}{c_0}\right) \simeq 1$ for all configurations i and j. Kirkpatrick et al. (1982) proposed the following empirical rule: choose a large value for c_0 and perform a fixed number of transitions. If the *acceptance coefficient* χ, defined as the number of approved transitions on the number of total transitions, is smaller than a fixed value χ_0, then c_0 is doubled. Such a procedure continues until χ exceeds χ_0. Such a rule has been refined by Johnson et al. (1991) that determines c_0 by calculating for a random number of transitions the average increase of the cost function. Therefore, the initial value of the control parameter is calculated as

$$c_0 = \frac{|\Delta C|}{\ln(\chi_0^{-1})}.$$

6.4.3 The Final Value of the Control Parameter

To determine the final value of the control parameter a *stop criterion* is defined. Such a criterion can be established in different ways. We give two examples. One consists in fixing the number K of transitions for which the algorithm must be performed. For all values $k = 1,\dots,K$ of the transitions, a value of the control parameter is calculated. Generally, K varies from 6 (Nahar et al., 1986) to 50 (Bonomi and Lutton, 1984). For instance, if $K = 8$, the final value of c_k is obtained after eight decrements of the same parameter. Another method proposed foresees finishing the algorithm if the objective function values are identical for a fixed number, say ρ, of consecutive Markov chains (Morgenstern and Shapiro, 1986).

6.4.4 Decrement of the Control Parameter

The decrement rule of the control parameter is chosen to establish a quasi-equilibrium for the fixed length of Markov chain. In the thermodynamical process of annealing it is very important that the temperature of the solid is decremented slowly in order to re-establish quasi-equilibrium for all the energetic stadium. The parallelism between the physical process and the mathematical process involves a decrement rule that allows only small variations of the control parameter value. Frequently, the variation of c_k value is accomplished by means of the following rule

$$c_{k+1} = \alpha \cdot c_k, \;\; k = 0,\dots,$$

where α is a coefficient close to 1.

Kirkpatrick et al. (1982) suggested for such a coefficient a value equal to 0.95, while Sechen and Sangiovanni-Vincentelli (1985) used values that vary from 0.5 to 0.99. Nahar et al. (1985) fixed the number of decrements K and the value c_k experimentally determined for all $k = 1,\dots,K$. Skishim and Golden (1983) and Golden and Skishim (1986) divided the interval $[0, c_0]$ into a preset number of K subintervals and c_k is calculated as

$$c_k = \frac{K-k}{K} \cdot c_0, \;\; k = 1,\dots,K.$$

Fig. 6.2 shows a control parameter trend. Starting from c_0, the temperature is maintained constant for L consecutive steps (length of the Markov chains). Then, after each series of L steps, the control parameter value is decremented by a fixed factor α, with $0 < \alpha < 1$.

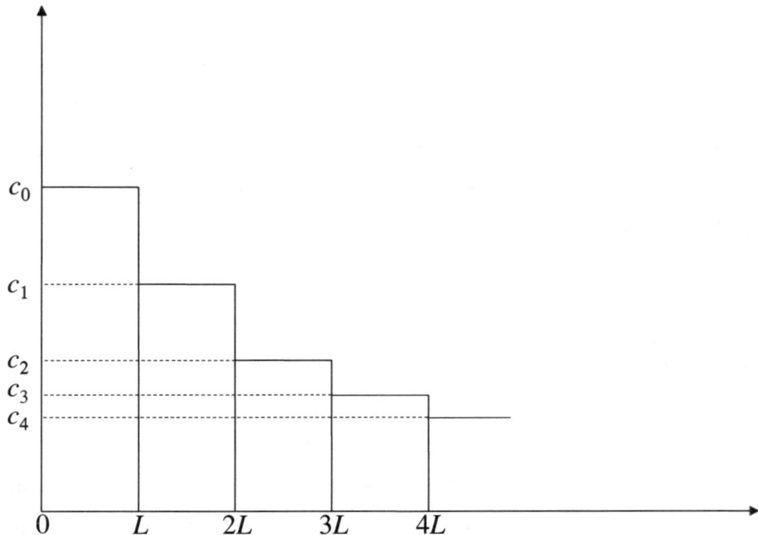

Fig. 6.2 The most common control parameter trend

What has been defined in the preceding subsections can be summarized in this general scheme of a SA algorithm.

Let s_0 be an initial solution;
Let $c_0 > 0$ be the initial value of the control parameter;
while (stopping criterion is not true){
    while (equilibrium criterion is not true) {
        Change configuration from($s_i \rightarrow s_j$)
        If ($\Delta C_{ij} < 0$) or $\left(\exp\left(-\frac{\Delta C_{ij}}{c_k}\right) > \text{random}[0,1]\right)$
            Update the current solution;
    }
    Update the control parameter.
}

6.5 Our Simulated Annealing Algorithm

This section describes how we have engineered our simulated annealing algorithm. We start by showing the procedure used to determine an initial schedule. Let

- t_{1e}: The starting time of a working shift for an employee e;
- t_{2e}: The finishing time of a working shift for an employee e;
- $t_{2\max}$: The maximum among the finishing times of the working shifts assigned until the current iteration;
- $t_{1\max}$: The maximum starting time of a working shift;
- $n_{\max}$: The maximum number of working shifts in the planning horizon T.

Algorithm initial solution

Step 0. Let $t_{2\max} = 0$; $t_{1e} = 0$ and $t_{2e} = 0$ $\forall e \in E$;
Step 1. Choose at random a number of employees to be scheduled and for each employee e in this set let $t_{1,e} = t_{2\max}$. Let t_{2e} be the finishing time of the shift. If $t_{2e} > t_{2\max}$ then $t_{2\max} = t_{2,e}$;
Step 2. If all the employees have been scheduled then go to **Step 3**, otherwise go back to **Step 1**;
Step 3. For each employee e choose the starting time $t_{1,e} > t_{2\max}$ respecting the minimum and maximum idle time allowed between two consecutive shifts; let $t_{2,e}$ be the finishing time of the shift. If $t_{2,e} > T$ set $t_{2,e} = T$;
Step 4. Let $t_{2\max} = \max_e\{t_{2,e}\}$; if $t_{2\max} \geq T$ or the number of assigned shifts for each employee is greater than or equal to $n_{\max}$ then **stop**, otherwise go back to **Step 3.**

6.5.1 Procedure for Determining the First Value of the Control Parameter

The first value of the control parameter is computed as follows.
Initialization
Define with s_0 the starting schedule and let C_0 be its cost; let cnt_conf be the number of accepted configurations for each value of the control parameter, and let cnt be the

number of generated configurations. In addition, let $\chi = 0$ be the initial value of the acceptance coefficient introduced in Sect. 6.4.1.2.

Generic iteration

Step 1. Apply a function to change configuration from s_i to s_j;

Step 2. Compute the cost C_j for the configuration s_j;

Step 3. If $\Delta C_{i,j} < 0$ or $\exp\left(-\frac{\Delta C_{ij}}{c_k}\right) > \text{random}[0,1]$ then set $cnt\_conf = cnt\_conf + 1$ and $s_i = s_j$, otherwise go back to **Step 1**;

Step 4. Let $\chi = \frac{cnt\_conf}{cnt}$ and $cnt = cnt + 1$;

Step 5. If $\chi > \chi_0$ then **stop**; otherwise $c_k = 2c_{k-1}$, $k = k+1$ and go back to **Step 1.**

Step 1 is accomplished by a function called *Perturb*($s_i \rightarrow s_j$) that is described in Sect. 6.5.4 and will be used also in the algorithm progresses to pass through different configurations.

6.5.2 Equilibrium Criterion and Stop Criterion

The equilibrium criterion controls the variation of the values assumed by c_k. Roughly speaking, after a fixed number of iterations the algorithm moves from c_k to c_{k+1}. The stop criterion, instead, determines the end of the simulated annealing algorithm and is as follows: when for two consecutive Markov chains any transition is no longer accepted then the stopping criterion is true and the algorithm ends.

6.5.3 Decrement Rule for the Control Parameter

Generally, the cooling rate of the control parameter depends on the objective function. For the examined problem we have selected a logarithmic decrement rule. Such a rule has been proposed by Hajek (1988). If we denote with c_k the control parameter value at the kth iteration, the decrement rule is

$$c_{k+1} = \frac{c_k}{\log(1+k)}.$$

The choice of such a rule has been dictated by the existence of a strong dichotomy between the choice of the control parameter and the choice of the Markov chain lengths.

6.5.4 *Function* Perturb*($s_i \rightarrow s_j$)*

The function *Perturb*($s_i \rightarrow s_j$) allows the algorithm to go from a configuration s_i to a configuration s_j. Two different techniques of perturbation are used to search for a new configuration. The first one, called *Shift*, operates by shifting to the left or to the right a working shift. The second one, called *Accord*, increases or decreases the duration of a working shift. At each iteration the algorithm chooses one between the two functions randomly. In the following, we report how such functions work.

Function *Shift*

Step 1. Choose an employee $e \in E$;
Step 2. Choose a shift i of e and let $t_{i,1e}$ and $t_{i,2e}$ be the starting time and the finishing time of the shift chosen, respectively;
Step 3. Calculate the idle hours hs to the left of $t_{i,1e}$ and the idle hours hd to the right of $t_{i,2e}$;
Step 4. Choose at random a number h of hours;
Step 5. If $hs > hd$ then move i to the left by h hours ($t_{i,1e} = t_{i,1e} - h$ and $t_{i,2e} = t_{i,2e} - h$) otherwise move i to the right by h hours ($t_{i,1e} = t_{i,1e} + h$ and $t_{i,2e} = t_{i,2e} + h$);
Step 6. If $t_{i,1e} > T$ then $t_{i,1e} = t_{1\max}$;
Step 7. If $t_{i,2e} > T$ then $t_{i,2e} = T$.

Function *Accord*

Step 1. Choose an employee $e \in E$;
Step 2. Choose a shift i of e and let $t_{i,1e}$ and $t_{i,2e}$ be the starting time and the finishing time of the shift chosen, respectively;
Step 3. Choose at random a number of hours $h < h_{\max}$;
Step 4. Let $num = random[0,10]$; if $num < 5$ and the shift duration is greater than h then shorten the shift i as $t_{i,2e} = t_{i,2e} - h$; if $num > 5$ go to **Step 5**;
Step 5. If $(t_{i-1,2e} - t_{i,1e}) > h$ then enlarge the shift i to the left by h hours ($t_{i,1e} = t_{i,1e} - h$); if $t_{i+1,1e} - t_{i,2e} > h$ then enlarge the shift i to the right by h hours ($t_{i,2e} = t_{i,2e} + h$);
Step 6. If $t_{i,1e} < 0$ set $t_{i,1e} = 0$ and if $t_{i,2e} > T$ set $t_{i,2e} = T$.

6.6 Experimental Results

6.6.1 Presentation of the Experiments and Implementation Details

The proposed simulated annealing algorithm has been implemented in the C language and tested on a Pentium IV PC with a 2.8 GHz processor and 512 MB RAM. We experimented with random instances with 30, 50, 100 and 500 employees, respectively. The workforce has been divided into five groups, each one representing a qualification, and the workforce scheduling has been carried out in a planning horizon $T = 168$ periods (a week divided into hour intervals) and a total number of shifts ranging from 6 to 8 for each employee. For each group of employees (30, 50, 100 and 500) and for each different initial value of the control parameter, 150 simulations have been executed. The initial values of the control parameter used in the experiments have been obtained by means of the procedure described in Sect. 4.1, and are shown in Table 6.2. Moreover, in Table 6.2 we also report the decrement rule used for the control parameter during the algorithm progresses. Table 6.3 shows the number of active employees used each time for each scenario. Table 6.4 shows the starting values of the weights w_e and w_i $(i = 1, \ldots, 6)$ associated with the objective function costs, as introduced in Sect. 6.3.2. Finally, Table 6.5 shows the values of the other characteristic parameters.

Table 6.2 Initial values of the control parameter used in the experiments

Initial values of the control parameter	*Decrement rule of the control parameter*
$c_0 = 6400$	$c_{k+1} = \frac{c_k}{\log(1+k)}$
$c_0 = 12,800$	$c_{k+1} = \frac{c_k}{\log(1+k)}$
$c_0 = 25,600$	$c_{k+1} = \frac{c_k}{\log(1+k)}$
$c_0 = 204,800$	$c_{k+1} = \frac{c_k}{\log(1+k)}$
$c_0 = 409,600$	$c_{k+1} = \frac{c_k}{\log(1+k)}$

In each scenario the characteristic parameters receive the same value as they are independent of the number of employees, while the number of active employees changes (see Table 6.3), due to its dependence on the total number of employees.

Table 6.3 Number of active employees each time for each scenario

Total number of employees	*Number of active employees each time*
30	8
50	15
100	30
500	60

Table 6.4 Starting values of the weights associated with the objective function costs used in the experiments

w_e	w_1	w_2	w_3	w_4	w_5	w_6
1	0.5	0.5	0.5	0.5	10	30

Table 6.5 Values of the characteristic parameters

Instance parameters	*Value*
Minimum duration of a shift	6
Maximum duration of a shift	8
Minimum working hours on the time horizon	40
Maximum working hours on the time horizon	48

6.6.2 Analysis of the Experiments

The algorithm behavior

The control parameter's initial value and its decrement rule influence the probability with which worse solutions are accepted. In fact, the acceptance probability of worse solutions depends on the Boltzmann distribution as follows:

$$\exp\left(\frac{-\Delta C_{ij}(k)}{c_k}\right) > \text{random}[0,1].$$

Therefore, higher values of c_k are associated with low probabilities to accept worse solutions (and viceversa), and thus a higher initial value of the control parameter tends to increase the number of iterations in which the acceptance probability is high.

Figures 6.3–6.7 show the algorithm behavior with respect to different values of c_0 and for 30, 50 and 100 employees. It should be noticed how the solutions trend depends on the initial values of the control parameter: the increase of c_0 results in a worse algorithm convergence, as well as in a larger number of iterations in which worse solutions are accepted.

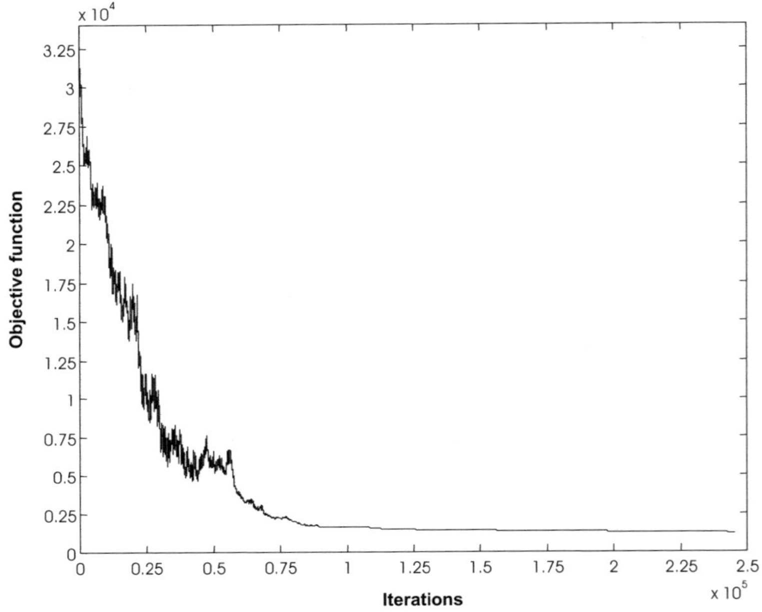

Fig. 6.3 The algorithm behavior for $c_0 = 6400$ and $|E| = 30$

The solution quality

Figures 6.8–6.11 report histograms showing the efficacy of the proposed algorithm. Each histogram analyzes the results obtained by the 150 runs of the algorithm on a given scenario. The classes (clusters) of the histograms are obtained by dividing the interval $0 - 100\%$ into 20 subintervals with equal width.

We measure the following percentage gap

$$\frac{TotalCost - L.B.Cost}{TotalCost}$$

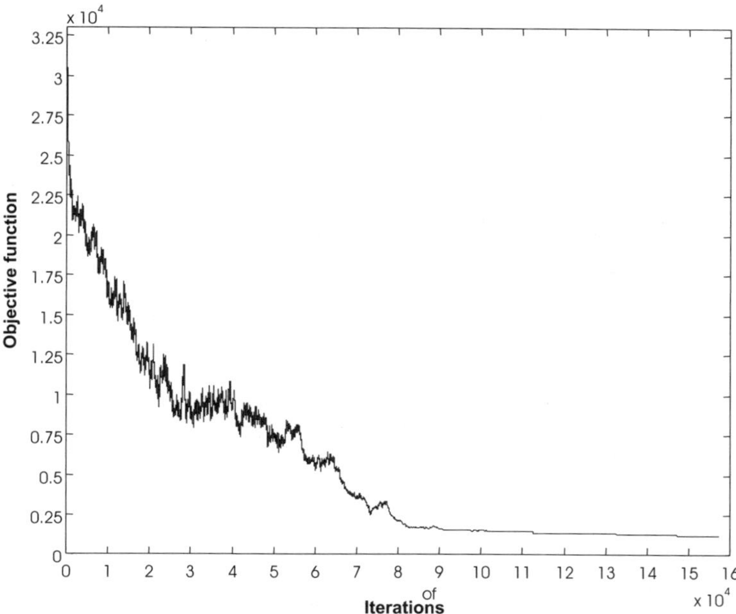

Fig. 6.4 The algorithm behavior for $c_0 = 12,800$ and $|E| = 30$

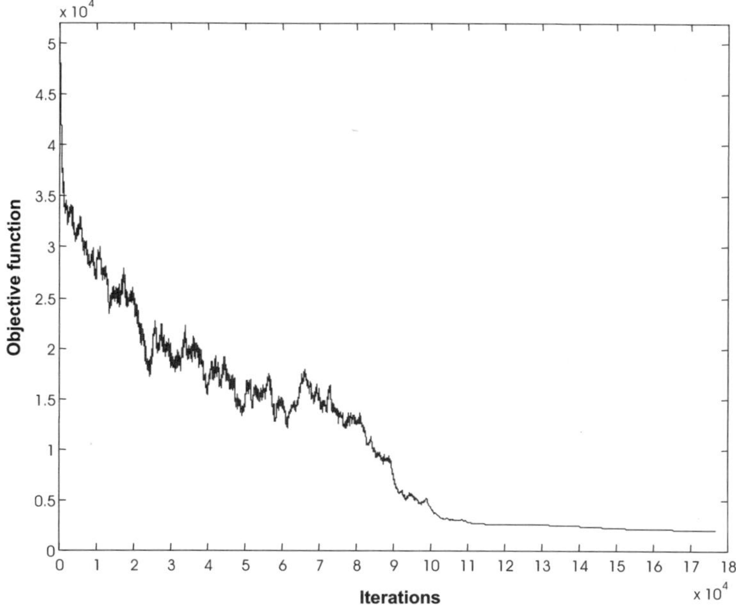

Fig. 6.5 The algorithm behavior for $c_0 = 25,600$ and $|E| = 50$

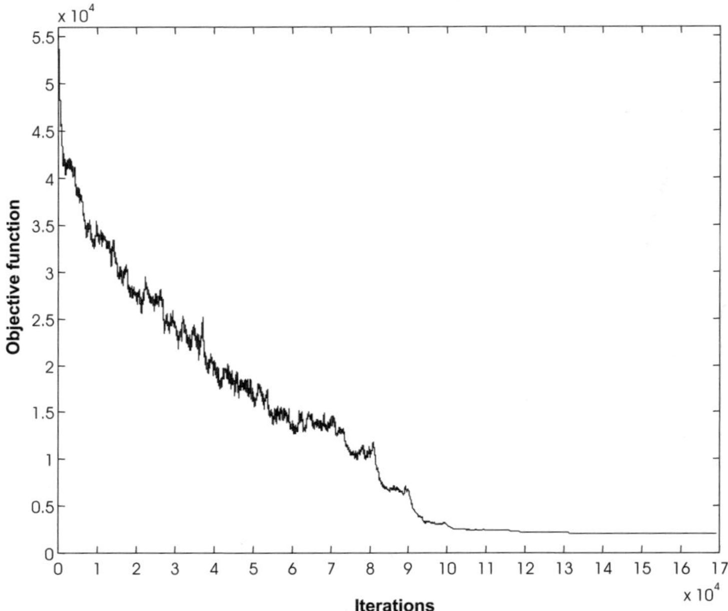

Fig. 6.6 The algorithm behavior for $c_0 = 204,800$ and $|E| = 50$

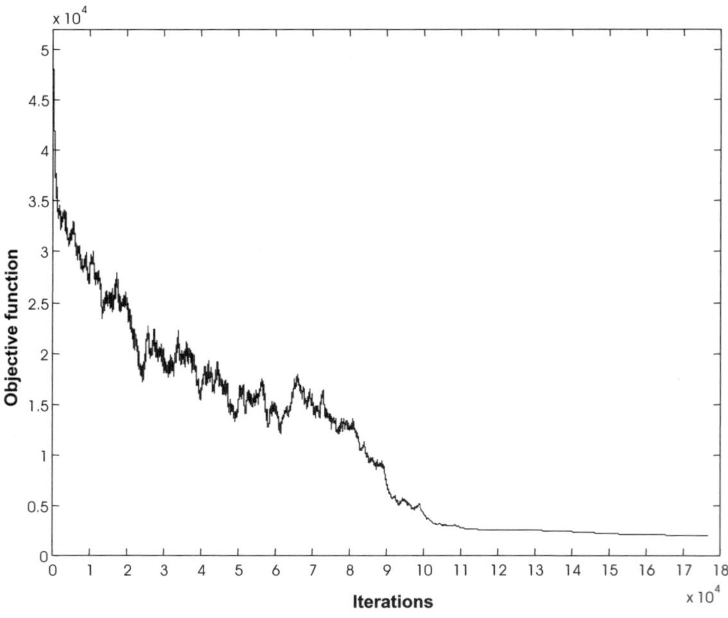

Fig. 6.7 The algorithm behavior for $c_0 = 409,600$ and $|E| = 100$

where *L.B. Cost* is a lower bound on the minimum schedule cost, given by the value of the objective function without penalty costs, while *Total Cost* is the best solution achieved by the algorithm.

The height of a rectangle associated with a class, measures the number of instances (out of the 150 tested for that scenario) in which the algorithm achieves a solution whose gap is less than or equal to that associated with the class. Consequently, all the solutions with a percentage gap less than 5% belong to class 1, all the solutions with a percentage gap in between 5% and 10% belong to class 2, and so on.

For the lowest value of c_0 (Fig. 6.8), the algorithm achieves a percentage gap of less than 5% for a number of times greater than 50 only for the scenario with 30 employees. In particular, for such a scenario an average gap less than 5% has been obtained 90 times. For the other scenarios, a percentage gap less than 5% has been achieved less than 40 times.

For higher values of the number of employees and for the lowest value of c_0, the percentage gap dispersion increases, while for higher values of c_0 the percentage gap dispersion decreases. The maximum average gap is obtained for the lowest values of the control parameter ($c_0 = 12,800$ and $c_0 = 51,200$) and for the scenarios with $|E| = 30$ and $|E| = 50$. The numbers of percentage gaps equal to zero are shown in Table 6.6 for all the values c_0 and for each scenario.

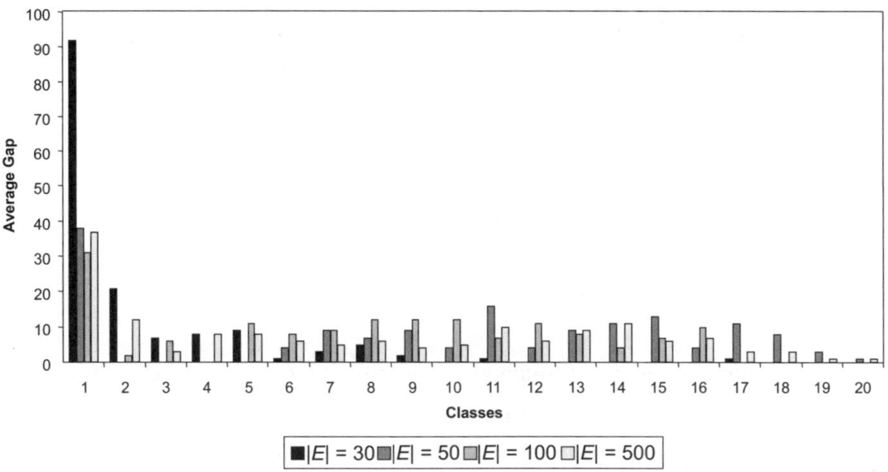

Fig. 6.8 Percentage gaps achieved in each scenario with $c_0 = 12,800$

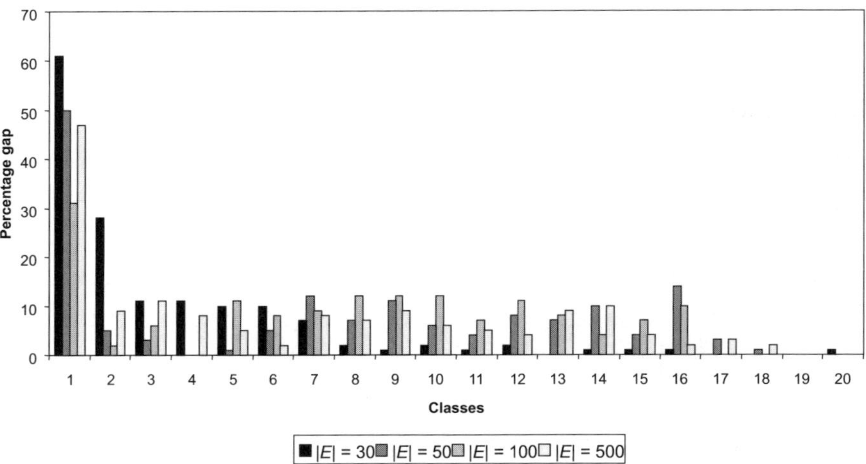

Fig. 6.9 Percentage gaps achieved in each scenario with $c_0 = 51,200$

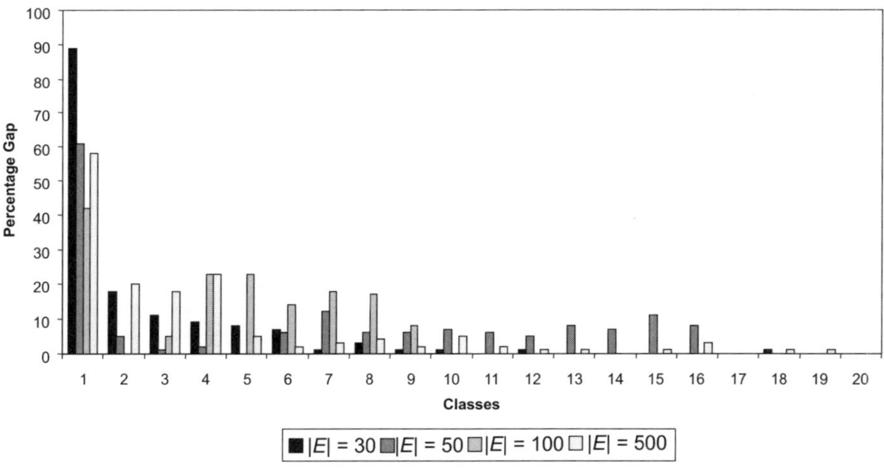

Fig. 6.10 Percentage gaps achieved in each scenario with $c_0 = 102,400$

6.6.3 Computational Times

In the following, we report our findings on the running times of the algorithm. Table 6.7 shows the maximum, minimum and average computational times for the four scenarios previously described.

From Table 6.7 we obtained the chart shown in Fig. 6.12, from which a nearly exponential trend with the number of employees can be observed.

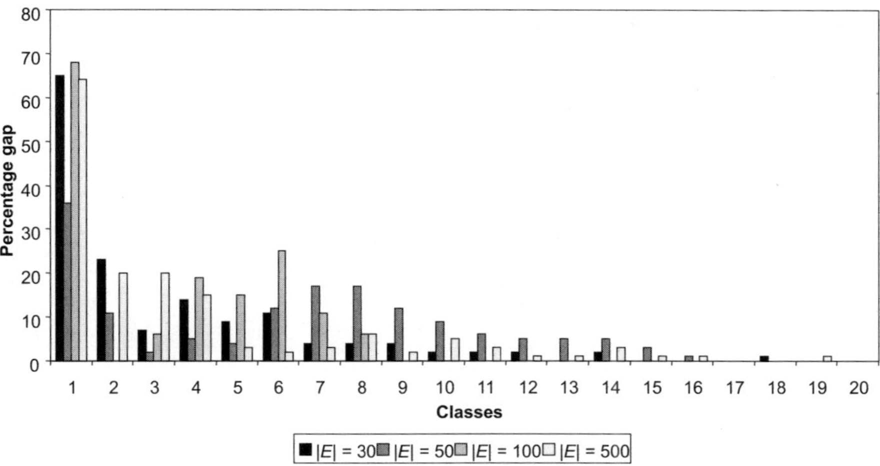

Fig. 6.11 Percentage gaps achieved in each scenario with $c_0 = 204,800$

6.7 Conclusions

In this chapter, we proposed a simulated annealing algorithm for a particular kind of staff-scheduling problem, known as the constrained heterogeneous workforce

Table 6.6 Percentages of times a gap equal to zero occurs in each scenario and for all the control parameter initial values

Number of employees	*Initial value of the control parameter*	*Percentage of times a gap equal to zero occurs*
$\vert E\vert = 30$	12,800	26%
$\vert E\vert = 30$	51,200	35.3%
$\vert E\vert = 30$	102,400	40.7%
$\vert E\vert = 30$	204,800	40%
$\vert E\vert = 50$	12,800	26%
$\vert E\vert = 50$	51,200	32%
$\vert E\vert = 50$	102,400	36.7%
$\vert E\vert = 50$	204,800	8%
$\vert E\vert = 100$	12,800	29.3%
$\vert E\vert = 100$	51,200	14.7%
$\vert E\vert = 100$	102,400	28%
$\vert E\vert = 100$	204,800	45.3%
$\vert E\vert = 500$	12,800	23.3%
$\vert E\vert = 500$	51,200	19.3%
$\vert E\vert = 500$	102,400	38%
$\vert E\vert = 500$	204,800	34.7%

Table 6.7 Minimum, maximum and average computational times associated with the analyzed scenarios

Number of employees	*Max CPU time (s)*	*Min CPU time (s)*	*Average CPU time (s)*
30	235.74	43.6	112.74
50	504.60	71.16	283.27
100	1797.05	908.15	1208.47
500	11,593.30	6501.51	9113.68

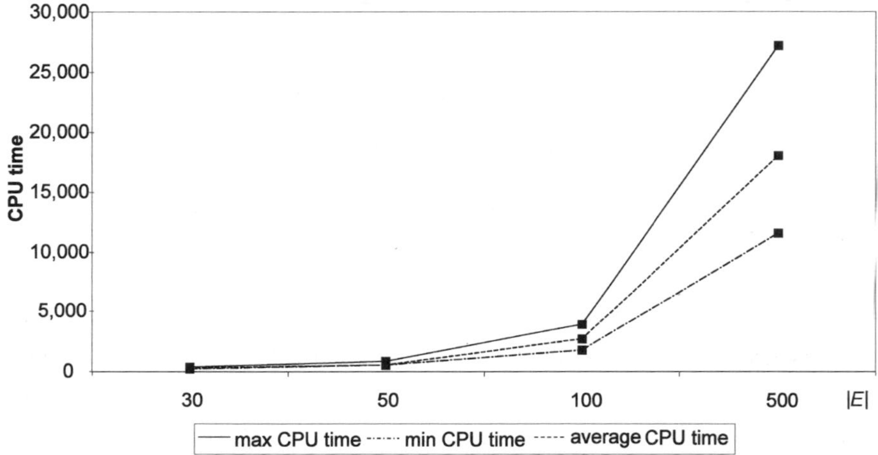

Fig. 6.12 Trend of the maximum, minimum and average CPU time for the analyzed scenarios

scheduling problem. The algorithm has been tested on various scenarios ranging from 30 to 500 employees and different degrees of freedom for the employees. We show the efficacy of the proposed method in finding good solutions in limited computational time compared with the difficulty of the problem.

Appendix A
AMPL Code: The Container-allocation Problem

This appendix contains an implementation in the AMPL language of the container-allocation problem presented in Sect. 3.3.

```
set BERTH;
set TERMINAL;
set SHIP;
set INSP_AREA;

param CAPACITY{TERMINAL};
param DISTANCE{BERTH,TERMINAL};
param DISTANCE2{TERMINAL,INSP_AREA};
param I{SHIP};
param NI{SHIP};
param W_I{b in BERTH, t in TERMINAL};
param W_NI{b in BERTH, t in TERMINAL};

var S{k in SHIP} binary;
var X_I{b in BERTH, t in TERMINAL} >= 0, integer;
var X_NI{b in BERTH, t in TERMINAL} >= 0, integer;
var Y{k in SHIP, b in BERTH} binary;

minimize cost: sum{b in BERTH, t in TERMINAL} W_I[b,t]*X_I[b,t] + sum{b
in BERTH, t in TERMINAL} W_NI[b,t]*X_NI[b,t] + sum{k in SHIP} S[k]*10000;
```

```
subject to CONTAINER_TO_BE_INSPECTED {b in BERTH}:
sum{t in TERMINAL} X_I[b,t] = sum{k in SHIP} I[k]*Y[k,b];

subject to CONTAINER_NOT_TO_BE_INSPECTED{b in BERTH}:
sum{t in TERMINAL} X_NI[b,t] = sum{k in SHIP} NI[k]*Y[k,b];

subject to BERTH_CAPACITYb in BERTH:
sum{k in SHIP} Y[k,b] <= 1;

subject to SHIP_CAPACITY {k in SHIP}:
sum{b in BERTH} Y[k,b] + S[k] = 1;
```

Appendix B
AMPL Code: The Inspection Scheduling Problem

This appendix contains an implementation in the AMPL language of the inspection scheduling problem presented in Sect. 3.4.

```
set TIME:=1,..,T;

param T;
param C_ISP {TIME};
param CONT_CONS {TIME};

var X{t in TIME}>=0, integer;
var S{t in TIME}>=0, integer;
var V{t in TIME}>=0, integer;

minimize NUMBER_MOVES:
sum{t in TIME} (α*V[t] + (1 − α)*S[t]);

subject to SAT_DELIVERY {k in TIME}:
sum{t in 1..k} (X[t] - V[t] + S[t]) = sum{t in 1..k} CONT_CONS[t];

subject to CAP_ISP{t in TIME}:
X[t] <= C_ISP[t];
```

Appendix C
Code in the C Language: The Iterative Penalty Method Algorithm

This appendix contains an implementation in the C language of the iterative penalty method (see Sect. 4.4), an algorithm that calculates a set of alternative paths, based on the Dijkstra algorithm for calculating each path.

```
#include <stdio.h>
#include <time.h>
#define mmax 5000 /*maximum number of arcs*/
#define nmax 3000 /*maximum number of nodes*/

typedef int vector1[nmax];
typedef int vector2[mmax];
typedef float vector3[mmax];
typedef float vector4[nmax];

/*definition of global variables*/
vector1 fs,p,q;
vector2 a;
vector3 w,w1;
vector4 l,l1,f;
int r,last,n,m,u,dest;
extern void readfstar();
extern void initialization();
extern void extract_min();
```

```
double exetime ();

void main() /*A (see the legend at the end of IPM algorithm description)*/
{
    FILE *out,*pat,*sim,*sim2,*out1,*sim3;
    int j,k,prec,temp,count,node_count,max;
    float minpath,c,b;
    double t1, t2;

    t1=exetime();
    count=1;
    max=0;
    readfstar();
    for (j=1;j<=m;j++)
    {
        w1[j]=w[j];
    }

    printf("insert the origin node: ");
    scanf("%d",&r);
    printf("insert the destination node: ");
    scanf("%d",&dest);
    printf("insert the maximum path length"); /*B (see the legend)*/
    scanf("%f",&c);
    printf("insert the arc penalty factor "); /*C (see the legend)*/
    scanf("%f",&b);

    if ((out=fopen("Global.txt","w"))==NULL) /*D (see the legend)*/
        printf("File opening error");

    if ((out1=fopen("length.txt","w"))==NULL) /*E (see the legend)*/
        printf("File opening error");

    if ((pat=fopen("paths1.txt","w"))==NULL) /*F (see the legend)*/
        printf("File opening error");
```

```
if ((sim=fopen("indexes.txt","w"))==NULL) /*G (see the legend)*/
    printf("File opening error");

if ((sim2=fopen("similar.txt","w"))==NULL) /* H-M (see the legend)
    printf("File opening error");

if ((sim3=fopen("copy.txt","w"))==NULL) /*L-M (see the legend)*/
    printf("File opening error");

initialization();
do /*N (see the legend)*/
{
    extract_min();
    if (fs[u]<fs[u+1])
    {
        for (j=fs[u];j<=(fs[u+1]-1);j++)
        {
            k=a[j];
            if (l[u]+w1[j]<l[k])
            {
                l[k]=l[u]+w1[j];
                l1[k]=l1[u]+w[j];
                p[k]=u;
                if (q[k]==0)
                {
                    q[last]=k;
                    last=k;
                    q[k]=(n+1);
                }
            }
        }
    }
    if (u==dest)
        if (count==1)
```

```
        minpath=l[u];
if (l1[dest]>(c*minpath)) /*P (see the legend)*/
    go to terminate;

fprintf(out,"Path number %d",count);
prec=u;
node_count=1;
fprintf(sim2,"%d %d",count,prec);
fprintf(sim3,"%d %d",count,prec);
temp=p[prec];
if (fs[temp]<fs[temp+1])
{
    for (j=fs[temp];j<=(fs[temp+1]-1);j++)
    {
        if (a[j]==prec)
        {
            fprintf(out,"from=%d to=%d weight=%f",p[prec],prec,w[j]);
            fprintf(pat,"s%d %d %d %d %f %f",count,count,p[prec],
                prec,w[j],l1[prec]);
            fprintf(sim2," %d",p[prec]);
            fprintf(sim3,"%d",p[prec]);
            w1[j]=w1[j]+(b*w1[j]);
        }
    }
}
prec=p[u];
node_count++;
while (prec!=r)
{
    temp=p[prec];
    if (fs[temp]<fs[temp+1])
    {
        for (j=fs[temp];j<=(fs[temp+1]-1);j++)
        {
            if (a[j]==prec)
```

```
                {
                    fprintf(out,"from=%d to=%d weight=%f ",p[prec],prec,w[j]);
                    fprintf(pat,"s%d %d %d %d %f %f ",count,count,p[prec],
                        prec,w[j],l1[prec]);
                    fprintf(sim2," %d",p[prec]);
                    fprintf(sim3," %d",p[prec]);
                    w1[j]=w1[j]+(b*w1[j]);
                }
            }
        }
        prec=p[prec];
        node_count++;;
    }
    fprintf(sim2," 0");
    fprintf(sim3,"" 0"); /*Q (see legend)*/

    if (node_count>max)
    {
        max=node_count;
        fprintf(out,"origin= %d destination= %d length=%f ",r,dest,l1[dest]);
        fprintf(out1,"%d %.3f ",count,l1[dest]); /*R (see the legend)*/
        fprintf(sim,"%d ",count);
        count++;
        initialization(); /*S (see the legend)*/
    }    }     while (last!=n+1);
terminate:fprintf(out,"=====End Shortest Path=====")

count−−; /*SP (see the legend)*/
fprintf(out,"The longest path has %d nodes n",max);
fprintf(out,"The generated alternative paths are %d ",count);
fclose(pat);
fclose(sim);
fclose(sim2);
fclose(sim3);
t2=exetime();
```

```
    fprintf(out,"Algorithm execution time:");
    fprintf(out,"%lf secondi ",t2-t1);
}
```

```
void readfstar(void) /*T (see the legend)*/
{
    int i;
    FILE *ifp;

    ifp=fopen("fstar.txt","r"); /* V (see the legend)*/
    fscanf(ifp,"%d",&n);
    fscanf(ifp,"%d",&m);
    for(i=1;i<=m;i++)
        fscanf(ifp,%d ",&a[i]);
    for(i=1;i<=m;i++)
        fscanf(ifp,%f",&w[i]);
    for(i=1;i<=(n+1);i++)
        fscanf(ifp,"%d ",&fs[i]);
    fclose(ifp);
}
```

```
void initialization(void) /*W (see the legend)*/
{
    int i;

    for (i=1;i<=n;i++)
    {
        l[i]=99999;
        l1[i]=99999;
        p[i]=0;
        q[i]=0;
```

```
    }
    l[r]=0;
    l1[r]=0;
    p[r]=r;
    q[n+1]=r;
    q[r]=n+1;
    last=r;
}
```

```
void extract_min(void) /*Y (see the legend)*/
{
    int x,i,min;
    min=3200;
    i=n+1;
    x=n+1;

    while ((q[i])!=(n+1))
    {
        if ((l[q[i]])<min)
        {
            x=i;
            min=(l[q[i]]);
        }
        i=q[i];
    }
    u=q[x];
    q[x]=q[u];
    q[u]=0;
    if (last==u)
        last=x;
}
```

```
double exetime (void) /*Z (see the legend)*/
{
    time_t ltime;
    double sec;

    if ((ltime=clock())==-1)
        printf("Execution time not available");
    sec=(double)ltime/CLK_TCK;
    return sec;
}
```

LEGEND

- A: the function "main" builds the alternative paths and prints the output.
- B: the paths, whose lengths are higher than a fixed threshold, are deleted to prevent their selection.
- C: it assumes that on the arcs a multiplicative factor of the penalty is applied.
- D: "Global.txt" provides the output: each path is described as a set of arcs with their respective weights.
- E: "length.txt" provides the path index with its length in order to calculate the average length of the paths.
- F: "paths.txt" provides the number of the path, and its arcs and respective weights.
- G: "indexes.txt" provides the indexes that identify the generated paths.
- H: "similar.txt" provides each path with its nodes written by row (to be sent to the function that deletes duplications).
- L: copy of "similar.txt".
- M: on the files "similar.txt" and "copy.txt", a "0" is inserted to the row corresponding to the end of the path.
- N: exploration of "u" forward star; procedure for determining the shortest path from the origin to the destination.

- P: thanks to this condition it detects a path longer than the shortest one of a quantity greater than a fixed threshold.
- Q: this instruction allows the identification of the path with the maximum number of nodes.
- R: on the file "length.txt" it inserts the path index and its length.
- S: it recalls the vector initialization procedure to find the next path.
- SP: the longest path length is inserted in order to size the character strings in the file that deletes path duplications.
- T: the procedure "fstar" reads the "fstar.txt" content.
- V: on the file "fstar.txt" the function reads in input the graph forward star.
- W: the "initialization" procedure provides the initialization of the:
  - predecessor vector "p",
  - vector "q" that contains the nodes candidate to enter in the shortest-path tree,
  - vector "l" that contains the current shortest paths (penalized) lengths,
  - vector "l1" that contains the shortest paths original lengths.
- Y: the procedure "extract_min" considers, among the nodes of vector "q", the one with the smallest label "l". Vector "q" is managed as a priority queue.
- Z: this function calculates the programme execution time.

Appendix D
Code in the C Language: The *P*-Dispersion Algorithm

This appendix contains an implementation of the algorithm that calculates a set of paths such that the minimal dissimilarity among them is the largest possible (see Sect. 4.4); the algorithm is based on two heuristic methods.

```
#include <stdio.h>
#include <stdlib.h>
#include <time.h>
#define nmax 200 /*maximum number of initial alternative paths*/
#define pmax 350 /*maximum size of "support" vector*/
#define maxp 30 /*maximum number of required dissimilar paths*/

typedef float matrix[nmax][nmax];
typedef int vet1[maxp];
typedef int vet2[nmax];
typedef float vet3[pmax];

/*definition of global variables*/
int n,p,dim,sel,row,col; /*A (see the legend)*/
matrix D; /*AA (see the legend)*/
FILE *out;

void read_matrix(void);
float min_vector(float a[pmax]);
```

```
int equal(vet1 Z,vet1 R);
double exetime();
int rand();

void main(void) /*B (see the legend)*/
{
    vet1 S,SS,S1,S2; /*C (see the legend)*/
    vet2 V,Ssel;
    vet3 App1,App2;
    int r,x1,x2,xk,i,j,q,w,z,result;
    float fsi,fsf,fss,fs1,fs2,min,app,sum;
    double t1,t2;

    t1=exetime();
    printf("P-Dispersion Algorithm");
    printf("Please insert the size of subset S to be generated ");
    scanf("%d",&p);
    printf("Insert the size of the dissimilarity matrix");
    scanf("%d",&n);
    printf("Insert the size of the swap vectors");
    scanf("%d",&dim);
    read_matrix();
    if((out=fopen("solution.txt","w"))==NULL)
        fprintf(out,"Opening output file error");

/*D (see the legend)*/
    for (r=1;r<=n;r++) /*E (see the legend)*/
        V[r]=0; /*F (see the legend)*/
    sel=n;
    while (sel>p) /*G (see the legend)*/
    {
        z=1;
        for(r=1;r<=n;r++)
        {
            if((V[r]==0)(z<=sel))
```

```
            Ssel[z]=r;
        z++;
    }
}
min=10.0;
i=1;
while (i<sel)
{
    for(j=i+1;j<=sel;j++)
    {
        if(Ssel[i]<Ssel[j])
        {
            if(D[Ssel[j]][Ssel[i]]<min)
            {
                min=D[Ssel[j]][Ssel[i]];
                x1=Ssel[j];
                x2=Ssel[i];
            }
        }
        else
        {
            if((D[Ssel[i]][Ssel[j]])<min)
            {
                min=D[Ssel[i]][Ssel[j]];
                x1=Ssel[i];
                x2=Ssel[j];
            }
        }
    }
    i++;
}
app=(rand()/32768.0); /*H (see the legend)*/
if(app>=0.5)
    xk=x1;
else
```

```
        xk=x2;
    V[xk]=1; /*L (see the legend)*/
    sel=sel-1;
}
    z=1;
    if(sel==p)
    {
        for (r=1;r<=n;r++)
        {
            if((V[r]==0)(z<=p))
            {
                S[z]=r;
                z++;
            }
        }
    }
    for (z=1;z<=pmax;z++) /*M (see the legend)*/
    {
        App1[z]=10.0;
        App2[z]=10.0;
    }
    do /*N (see the legend)*/
    {
        for (w=1;w<=p;w++)
        {
            fprintf(out,"S[%d]=%d ",w,S[w]);
            min=10.0;
            i=1;
        }
        while (i<p)
        {
            for(j=(i+1);j<=p;j++)
            {
                if(S[i]<S[j])
                {
```

```
            if(D[S[j]][S[i]]<min)
            {
                min=D[S[j]][S[i]];
                x1=S[j];
                x2=S[i];
            }
        }
        else
        {
            if((D[S[i]][S[j]])<min)
            {
                min=D[S[i]][S[j]];
                x1=S[i];
                x2=S[j];
            }
        }
    }
    i++;
}
fsi=min;
fprintf(out,"f(s)_iniziale=%f ",fsi); /*P (see the legend)*/
for (w=1;w<=p;w++)
{
    SS[w]=S[w];
    fss=fsi;
    for(w=1;w<=p;w++)
    {
        S1[w]=S[w];
        S2[w]=S[w];
    }
    r=1; /*Q (see the legend)*/
    while (r<=n)
    {
        if(V[r]==1)
        {
```

```
                    xk=r; /*R (see the legend)*/
                    for (q=1;q<=p;q++)
                    {
                        if(S[q]==x1)
                            S1[q]=xk; /*S (see the legend)*/
                        for (q=1;q<=p;q++)
                        {
if(S[q]==x2)
    S2[q]=xk; /*SR (see the legend)*/
i=1;
z=1;
while (i<p)
{
    for (j=i+1;j<=p;j++)
    {
        if(S1[i]<S1[j])
            App1[z]=D[S1[j]][S1[i]];
        else
            App1[z]=D[S1[i]][S1[j]];
        if(S2[i]<S2[j])
            App2[z]=D[S2[j]][S2[i]];
        else
            App2[z]=D[S2[i]][S2[j]];
        z++;
    }
    i++;
}
fs1=min_vector(App1);
fs2=min_vector(App2);
if((fs1>fss)(fs1>=fs2)) /* T (see the legend)*/
{
    for (w=1;w<=p;w++)
    {
        SS[w]=S1[w];
        fss=fs1;
```

```
        }
        if((fs2>fss)(fs2>fs1)) /* TS (see the legend)*/
        {
            for (w=1;w<=p;w++)
            {
                SS[w]=S2[w];
                fss=fs2;
            }
        }/* End if V[r]==1 */
        r++;
    }
    result=equal(S,SS);
    if(result==0) /*V (see the legend)*/
    {
        for(w=1;w<=p;w++)
            V[S[w]]=1;
        for(w=1;w<=p;w++)
            V[SS[w]]=0;
        for(w=1;w<=p;w++)
            S[w]=SS[w];
    }
    else
    {
        fsf=fss;
        for (w=1;w<=p;w++)
        {
            fprintf(out,"S[%d]=%d ",w,S[w]);
            fprintf(out," Minimum dissimilarity in S=%f",fsf);
            sum=0;
            i=1;
            while (i<p)
            {
                for (j=i+1;j<=p;j++)
                    if(S[i]<S[j])
                        sum=sum+D[S[j]][S[i]];
```

```
                    else
                        sum=sum+D[S[i]][S[j]];
                i++;
            }
            fprintf(out,"Average dissimilarity in S=%f",(sum/dim));
        }
    }
    while (result==0); /*End do-while cycle*/
    t2=exetime();
    fprintf(out,"Algorithm execution time:");
    fprintf(out,"%lf secondi ",t2-t1);
}

/*W (see the legend)*/
```

```
void read_matrix(void) /*Y (see the legend)*/
{
    FILE *mat;
    if((mat=fopen("Matrix.txt","r"))==NULL) /*YY (see the legend)*/
    {
        fprintf(out,"Opening input file error");
        for (row=1;row<=n;row++)
        {
            for (col=1;col<=n;col++)
            fscanf(mat,"%f ",&D[row][col]);
    }
    fclose(mat);
}
```

```
float min_vector(float a[pmax]) /*Z (see the legend)*/
{
    int h;
```

```
    float minimum;

    minimum=10.0;
    for (h=1;h<=dim;h++)
        if(a[h]<minimum)
            minimum=a[h];
    return minimum;
}
```

```
int equal(vet1 Z,vet1 R) /*ZZ (see the legend)*/
{
    int ug,u;

    ug=1;
    for (u=1;u<=p;u++)
    {
        if(Z[u]!=R[u])
        {
            ug=0;
            break;
        }
    }
    return ug;
}
```

```
double exetime (void) /*ZZZ (see the legend)*/
{
    time_t ltime;
    double sec;

    if((ltime=clock())==-1)
```

```
        printf("Execution time not available");
    sec=(double)ltime/CLK_TCK;
    return sec;
}
```

LEGEND

- A: if p is even then $dim = p \cdot \lfloor (p-1)/2 \rfloor + p/2$;otherwise, i.e., p is odd, $dim = p \cdot \lfloor (p-1)/2 \rfloor$.
- AA: D is the "dissimilarity" matrix computed among all the pairs of alternative paths.
- B: the function "main" containing the two heuristics.
- C: S is the dissimilar path set to be improved.
- D: "Semi-greedy deletion heuristic" algorithm.
- E: all paths are in V.
- F: 0 means "selected path".
- G: the next cycle has to be repeated until selecting p paths in S.
- H: an element xk is chosen at random between $x1$ and $x2$, the nearest points in S.
- L: xk is excluded from S.
- M: "Pairwise interchange heuristic" algorithm.
- N: the next cycle has to be repeated till there are no changes in S.
- P: S is copied in SS to allow possible changes in S and, therefore, possible improvements of the initial objective function value fsi.
- Q: all points xk are changed with $x1$ and $x2$ in VnS, and the most favorable change is considered only, i.e., the one with the greatest fsi increase.
- R: xk is a generic point in VnS.
- S: $S1 = \{Sn\{x1\}\} \cup \{xk\}$.
- SR: $S2 = \{Sn\{x2\}\} \cup \{xk\}$.
- T: if there is an improvement, $S1$ is copied in S.
- TS: if there is an improvement, $S2$ is copied in S.
- V: it controls if $SS <> S$, that is if there were exchanges in S. SS is copied in S and the do-while cycle is repeated till there are not exchanges in S, that is $SS = S$.
- W: the algorithm ends because improvements are not possible in S.

- Y: the procedure reads in input the "dissimilarity" matrix; the matrix elements are stored by column.
- YY: "Matrix.txt" contains the dissimilarity matrix obtained by a GIS system that returns the output on the "solution.txt" file containing the desired subset of dissimilar paths.
- Z: the function returns the vector minimum element.
- ZZ: the function controls if two vectors are equal.
- ZZZ: the function calculates the programme execution time.

References

1. Aashtiani H.Z., The multi-modal traffic assignment problem, PhD Thesis. MIT (1979).
2. Abkowitz M., M. Lepofsky, P. Cheng, Selecting criteria for designating hazardous materials highway routes, Transportation Research Record 1333, pp. 30–35 (1992).
3. Ahuja K., T. Magnanti, J. Orlin, Network flows, Prentice Hall, New York (1993).
4. Akgün, V., E. Erkut, R. Batta, On finding dissimilar paths, European Journal of Operational Research 121, pp. 232–246 (2000).
5. Alicke K., Modeling and optimization of the intermodal terminal Mega Hub, OR Spectrum 24 (117), pp. 1–17 (2002).
6. Angel E., E. Bampis, L. Gourvès, Approximating the Pareto curve with local search for the bicriteria TSP(1,2) problem. Theoretical Computer Science 310, pp. 135-146 (2004).
7. Angel E., E. Bampis, L. Gourvès, J. Monnot, (Non-)approximability for the multi-criteria TSP(1,2), in Liskiewicz M., R. Reischuk (eds), Proc. of the 15th International Symposium on Fundamentals of Computation Theory, Lecture Notes in Computer Science 3623, pp. 329-340, Springer-Verlag, Berlin, Heidelberg (2005).
8. Arunapuram S., K. Mathur, D. Solow, Vehicle routing and scheduling with full truckloads, Tranportation Science 37 (2), pp. 170–182 (2003).
9. Audet C., G. Savard, W. Zghal, New branch-and-cut algorithm for bilevel linear programming, Journal of Optimization Theory and Applications 134, pp. 353–370 (2007).
10. Baldacci R., M. Battarra, D. Vigo, Routing a heterogeneous fleet of vehicles, Technical Report n.1, Dipartimento di Ingegneria Elettronica, Informatica e Sistemistica, OR. INGCE (2007).
11. Balobo F., S. Pinson, Dynamic modeling of a disturbance in a multiagent system for traffic regulation, Decision Support Systems 41 (1), pp. 131–146 (2005).
12. Bard J.F., J. Moore, J., A branch-and-bound algorithm for the bilevel programming problem, SIAM Journal of Scientific and Statistical Computing 11, pp. 281–292 (1990).
13. Barrett C., K. Bisset, M. Holzer, G. Konjevod, M. Marathe, Implementations of routing algorithms for transportation networks, Workshop on the DIMACS Shortest-Path Challenge (2006).
14. Bartholdi J.J., A guaranteed accuracy round-off algorithm for cyclic scheduling and set covering, Operations Research 29, pp. 501–510 (1981).

15. Batta R., S.S. Chiu, Optimal obnoxious paths on a network: transportation of hazardous materials, Operations Research 36 (1), pp. 84–92 (1988).
16. Bergkvist M., P. Davidsson, J.A. Persson, L. Ramstedt, A Hybrid micro-simulator for determining the effects of governmental control policies on transport chains, Multi-Agent-Based Simulation IV, Lecture Notes in Artificial Intelligence 3415, Springer-Verlag, Berlin, Heidelberg (2005).
17. Beroggi G., W. Wallace, Operational control of the transportation of hazardous materials: An assessment of alternative decision models, Management Science 41, pp. 1962–1977 (1995).
18. Bialas W., M. Karwan, Two-level linear programming, Management Science 30, pp. 1004–1020 (1984).
19. Bialas W., M. Karwan, J. Shaw, A parametric complementarity pivot approach for two-level linear programming, Operations Research Program Report 80-2, State University of New York at Buffalo (1980).
20. Boardman B., Intermodal transportation routing problem, Proc. of Portland International Conference on Management of Engineering and Technology. Portland, USA, p. 447 (1999).
21. Boffey T., B. Karzakis, Linear versus non linear models for hazmat routing, INFOR 33 (2), pp. 114–117 (1995).
22. Bonomi E., J.L. Lutton, The asymptotic behaviour of quadratic sum assignment problems: a statistical mechanics and the Metropolis algorithm, SIAM Review 26, pp. 551–568 (1984).
23. Bonomi E., J.L. Lutton, The asymptotic behaviour of quadratic sum assignment problems: a statistical mechanics approach, European Journal of Operational Research 26, pp. 295–300 (1986).
24. Bontekoning Y.M., C. Macharis, J.J. Trip, Is a new applied transportation research field emerging? A review of intermodal rail-truck freight transport literature, Journal of Transportation Research 38 (1), pp. 1–34 (2004).
25. Bonvicini S., P. Leonelli, G. Spadoni, Risk analysis of hazardous materials transportation: Evaluating uncertainty by means of fuzzy logic, Journal of Hazardous Materials 62, pp. 59–74 (1998).
26. Bouza G., G. Still, Mathematical programs with complementarity constraints: convergence properties of a smoothing method, Mathematics of Operations Research 32, pp. 467–483 (2007).
27. Brainard J., A. Lovett, J. Parfitt, Assessing hazardous waste transport risk using GIS, International Journal of Geographical Information System 10 (7), pp. 831–849 (1996).
28. Brogan J., J. Cashwell, Routing models for the transportation of hazardous materials-state level enahancements and modifications, Transportation Research Record 1020, pp. 19–22 (1985).
29. Brumbaugh-Smith J., D. Shier, An empirical investigation of some bicriterion shortest path algorithms, European Journal of Operational Research 43 (2), pp. 216–224 (1989).
30. Brusco M.J., L.W. Jacobs, A simulated annealing approach to the cyclic staff scheduling problem, Naval Research Logistics 40, pp. 69–84 (1993).
31. Buchheit M., N. Kuhn, J.P. Müller, M. Pischel, MARS: Modeling a multiagent scenario for shipping companies. European Simulation Symposium, Society for Computer Simulation (1992).

32. Bürckert H.-J., P. Funk, G. Vierke, An intercompany dispatch support system for intermodal transport chains, 33rd Hawaii International Conference on System Science (2000).
33. Cai X., K.N. Li, A genetic algorithm for scheduling staff of mixed skills under multi-criteria, European Journal of Operational Research 125 (2), pp. 359–369 (2000).
34. Candler W., R. Townsley, A linear two-level programming problem, Computers and Operations Research 9, pp. 59–76 (1982).
35. Caramia M., F. Guerriero, A heuristic approach to long-haul freight transportation with multiple objective functions, Omega, doi:10.1016/j.omega.2008.02.001.
36. Caramia, M., P. Dell'Olmo, M. Gentili, P.B. Mirchandani, Delivery Itineraries and Distribution Capacity of a Freight Network with Time Slots, Computers and Operations Research 34 (6), pp. 1585–1600 (2007).
37. Carrera S., K. Chami, R. Guimaraes, M.C. Portman, W. Ramdane Charif, Negotiation models for logistic platform planning and scheduling, Proceeding of the 11th Workshop on Project Management and Scheduling, pp. 43–47 (2008).
38. Chao I.M., A tabu search method for the truck and trailer routing problem, Computers and Operations Research 29, pp. 33–51 (2002).
39. Charnes A., W.W. Cooper, Management models and industrial applications of linear programming, John Wiley & Sons volumes I and II, New York (1961)
40. Charnes A., W.W. Cooper, R.O. Ferguson, Optimal estimation of executive compensation by linear programming, Management Science 1 (2), pp. 138–151 (1955).
41. Clímaco J.C.N., E.Q.V. Martins, A bicriterion shortest path algorithm, European Journal of Operational Research 11, pp. 399–404 (1982).
42. Chankong V., Y. Haimes, Multiobjective decision making theory and methodology, Elsevier Science, New York (1983).
43. Coello C.A., A comprehensive survey of evolutionary-based multiobjective optimization techniques, Knowledge and Information Systems. An International Journal 1 (3), pp. 269–308 (1999).
44. Cordeau J.F., M. Gendreau, G. Laporte, J.Y. Potvin, F. Semet, A guide to vehicle routing heuristics, Journal of the Operational Research Society 53, pp. 512–522 (2002).
45. Cormen T.H., C.E. Leiserson, R.L. Rivest, C. Stein, Introduction to algorithms, Second Edition. MIT Press and McGraw-Hill, London (2001).
46. Colson B., P. Marcotte, G. Savard, Bilevel programming: A survey, 4OR - Quarterly Journal of the Belgian, French and Italian Operations Research Societies 3, pp. 87–107 (2005).
47. Cox R., M. Turnquist, Scheduling truck shipments of hazardous materials in the presence of curfwews, Transportation Research Record 1063, pp. 21–26 (1986).
48. Crainic T.G., Long haul freight transportation, in Hall R.W. (ed.), Handbook of Transportation Science, 2nd Edition, Kluwer Academic Publishers, Amsterdam (2002).
49. Crainic T.G., J. Damay, M. Gendreau, An integrated freight transportation modelling framework, International Network Optimization Conference, Spa, Belgium (2007).
50. Current J., C. Revell, J. Cohon, The minimum covering shortest path problem, Decision Sciences 19, pp. 490–503 (1988).
51. Dafermos S., General Multimodal Network equilibrium problem with elastic demand, Networks 12 (1), pp. 57–72 (1982).

52. Davidsson P., L. Henesey, L. Ramstedt, J. Törnquist, F. Wernstedt, Agent-based approaches to transport logistics, AAMAS Workshop on Agents in Traffic and Transportation, New York (2004).
53. Deb K., Multi-objective optimization using evolutionary algorithms, Wiley, Chichester, UK (2001).
54. Deb K., A tutorial on evolutionary multi-objective optimization (EMO), Dagstuhl Seminar Proceedings 04461 Practical Approaches to Multi-Objective Optimization http://drops.dagstuhl.de/opus/volltexte/2005/252 (2005).
55. Dell'Olmo, P., M. Gentili, A. Scozzari. On finding dissimilar Pareto-Optimal paths, European Journal of Operational Research 162, pp. 70–82 (2005).
56. Dussault J.-P., P. Marcotte, S. Roch, G. Savard, A smoothing heuristic for a class of bilinear bilevel programs, European Journal of Operational Research 174 (3), pp. 1396–1413 (2006).
57. Easton F.F., N. Mansour, A distributed genetic algorithm for employee staffing and scheduling problems, Proceedings of the Fifth International Conference on genetic algorithms, pp. 360–367 (1993).
58. Ehrgott M., Approximation algorithms for combinatorial multicriteria optimization problems, International Transactions in Operational Research 7 (1), pp. 5-31 (2000).
59. Ehrgott M., Multicriteria optimization, Springer-Verlag, Berlin, Heidelberg (2005).
60. Ehrgott M., A discussion of scalarization techniques for multiple objective integer programming, Annals of Operational Research 147, pp. 343–360 (2006).
61. Ehrgott M., T.P. Dagmar, S. Thomas, A level set method for multiobjective combinatorial optimization: application to the quadratic assignment problem, Pacific Journal of Optimimization 2 (3), pp. 521–544 (2006).
62. Ehrgott M., X. Gandibleux, A survey and annotated bibliography of multiobjective combinatorial optimization, OR Spectrum 22 (4), pp. 425-460 (2000).
63. Ehrgott M., X. Gandibleux, Multiple criteria optimization: state of the art annotated bibliographic survey, Kluwer Academic Publishers, Boston (2002).
64. Erghott M., S. Rusika, An improved ε-constraint method for multiobjective programming, Report in Wirtschaftsmathematik (WIMA Report) (2005).
65. Ehrgott, M., A.J.V. Skriver, Solving biobjective combinatorial max-ordering problems by ranking methods and a two-phases approach, European Journal of Operational Research 147, pp. 657–664 (2003).
66. Ehrgott M., H.W. Hamacher, K. Klamroth, S. Nickel, A. Schöbel, M.M. Wiecek, A note on the equivalence of balance points and Pareto solutions in multiple-objective programming, Journal of Optimization Theory and Applications 92 (1), pp. 209–212 (1997).
67. Ewman A.M., C.A. Yano, Scheduling direct and indirect trains and containers in an intermodal setting, Journal of Transportation Science 34 (3), pp. 256–270 (2000).
68. Erkut E., F. Gzara, Solving the hazmat transport network design problem, Computers and Operations Research (35), pp. 2234–2247 (2008).
69. Erkut E., S.A. Tjandra, V. Verter, Hazardous materials transportation, in Barnhart C., G. Laporte (eds), Operations Research Handbook on Transportation, Elsevier, Amsterdam, pp. 539–621 (2007).

70. Erkut E., V. Verter, Modeling of transport risk for hazardous materials, Operations Research 46 (5), pp. 625–642 (1998).
71. Falk J.E., J. Liu, On bilevel programming, part I: general nonlinear cases, Mathematical Programming 70, pp. 47–72 (1995).
72. Fonseca C.M., P.J. Fleming, An overview of evolutionary algorithms in multiobjctive optimisation, Evolutionary Computation 3 (1), pp. 1–16 (1995).
73. Fukushima M., J.S. Pang, Complementarity constraint qualifications and simplified b-stationarity conditions for mathematical programs with equilibrium constraints, Computational Optimization and Applications 13, pp. 111–136 (1999).
74. Gambardella L.M., E. Taillard, G. Agazzi, MACS-VRPTW: A mutiple ant colony system for vehicles routing problems with time windows, in Corne D., M. Dorigo, F. Glover (eds), New Ideas in Optimization, McGraw Hill, London, pp. 63–76 (1999).
75. Gandibleux X., F. Beugnies, S. Randriamasy, Martins' algorithm revisited for multi-objective shortest path problems with a MaxMin cost function, 4OR - Quarterly Journal of the Belgian, French and Italian Operations Research Societies 4, pp. 47–59 (2006).
76. Gandibleux X., N. Mezdaoui, A. Frville, A multiobjective tabu search procedure to solve combinatorial optimization problems, in Caballero R., F. Ruiz, R. Steuer (eds), Advances in Multiple Objective and Goal Programming. Lecture Notes in Economics and Mathematical Systems 455, Springer Verlag Berlin Heidelberg, pp. 291–300 (1997).
77. Geoffrion A.M., Proper efficiency and the theory of vector maximization, Journal of Mathematical Analysis and Applications 22, pp. 618–630 (1968).
78. Glickman T.S., Rerouting Railroad Shipments of Hazardous Materials to Avoid Populated Areas, Accident Analysis and Prevention 15, pp. 329–335 (1983).
79. Golden B.L., C.C. Skishim, Using simulated annealing to solve routing and location problems, Naval Research Logistics Quarterly 33, pp. 261–279 (1986).
80. Gopalan R., R. Batta, M.H. Karwan, The equity constrained shortest path problem, Computers and Operations Research 17, 297–307 (1990a).
81. Gopalan R., K.S. Kolluri, R. Batta, M.H. Karwan, Modeling equity of risk in the transportation of hazardous materials, Operations Research 38, pp. 961–973 (1990b).
82. Ghiani G., G. Laporte, R. Musmanno, Introduction to Logistics Systems Planning and Control, John Wiley and Sons, New York, (2004).
83. Goldberg D.E., P. Segrest, Finite Markov chain analysis of genetic algorithms, in Grefenstette H.-M. (ed.), Genetic algorithms and their applications. Proceedings of the Second International Conference on Genetic Algorithms, Hillsdale, NJ, pp. 1–8 (1987).
84. Günther H.-O, K.H. Kim (eds), Container terminals and automated transport systems logistics control, issues and quantitative decision support, Spriger-Verlag, Berlin, Heidelberg (2005).
85. Gutin G., A.P. Punnel, The traveling salesman problem and its variations, Springer-Verlag, Berlin (2006).
86. Hajek B., Cooling schedules for optimal annealing, Mathematics of Operations Research 13, pp. 311–329 (1988).
87. Hajela P., C.Y. Lin, Genetic search strategies in multicriterion optimal design, Structural Optimization 4, pp. 99–107 (1992).

88. Hansen P., Bicriteria path problems, in Fandel G., T. Gal (eds), Multiple criteria decision making theory and applications, Lecture Notes in Economics and Mathematical Systems 177, pp. 109–127 Springer-Verlag, Berlin (1980).
89. Hansen M.P., Metaheuristics for multiple objective combinatorial optimization, Ph.D. Thesis, IMM-PHS-1998-45, Technical University of Denmark, Lyngby (1998).
90. Hansen P., Jaumard, B., Savard, G., New branch-and-bound rules for linear bilevel programming, SIAM Journal of Scientific and Statistical Computing (13), pp. 1194–1217 (1992).
91. Hartmann S., A general framework for scheduling equipment and manpower at container terminals, OR Spectrum 26 (1), pp. 51–74 (2004).
92. Henig M., The shortest path problem with two objective functions, European Journal of Operational Research 25, pp. 281–291 (1986).
93. Jaszkiewicz A., multiple objective metaheuristic algorithms for combinatorial optimization, Techical Report, Politechnika Poznanska, Poznan (2001).
94. Jarrah A.I.Z., J.F. Bard, A.H. De Silva, Solving large-scale tour scheduling problems, Management Science 40, pp. 1124–1150 (1994).
95. Jin H., R. Batta, M. Karwan, On the analysis of two new models for transporting hazardous materials, Operations Research 44, pp. 710–723 (1996).
96. Johnson D.S., C.R. Aragon, L.A. McGeoch, C. Schevon, Optimization by simulated annealing: an experimental evaluation; Part II, graph coloring and partitioning, Operations Research 39, pp. 378–406, (1991).
97. Johnson P.E., D.S. Joy, D.B. Clarke, J.M. Jacobi, HIGWAY 3.01, An enhanced highway routing model: Program, description, methodology and revised user's manual, Oak Ridge National Laboratory, ORLN/TM-12124, Oak Ridge, TN (1992).
98. Júdice J., A. Faustino, A sequential LCP method for bilevel linear programming, Annals of Operations Research 34, pp. 89–106 (1992).
99. kb78 Kalelkar A., R. Brooks, Use of multidimensional utility functions in hazardous shipment decisions, Accident Analysis and Prevention 10, pp. 251–265 (1978).
100. Kara B.Y., V. Verter, Designing a road network for hazardous materials transportation, Transportation Science 38 (2), 188–196 (2004).
101. Kefi M., O. Korbaa, K. Ghedira, P. Yim, Container handling using multi-agent architecture, in Lecture Notes in Computer Science 4496, Springer-Verlag, Berlin, Heidelberg (2007).
102. Kirkpatrick S., C.D. Gelatt Jr., M.P. Vecchi, Optimization by simulated annealing, IBM Research Report RC 9355 (1982).
103. Kocvara M., J.V. Outrata, Optimization problems with equilibrium constraints and their numerical solution, Mathematical Programming 101, pp. 119–149 (2004).
104. Kostreva M.M., M.M. Wiecek, Time dependency in multiple objective dynamic programming, Journal of Mathematical Analysis and Applications 173 (1), pp. 289–307 (1993).
105. Kuby M., X. Zhongyi, X. Xiaodong, A minimax method for finding the k best differentiated paths, Geographical Analysis 29 (4), pp. 298–313 (1997).
106. Jang Y., E. Kim, Study on the optimized model of intermodal transport using genetic algorithm, Shipping Logistics Research 45, pp. 75–98 (2005).

107. Laporte G., M. Gendreau, J. Potvin, F. Semet, Classical and Modern Heuristics for the Vehicle Routing problem, International Transations in Operational Research 7, pp. 285–300 (2000).
108. Lepofsky M., M. Abkowitz, P. Cheng, Transportation hazard analysis in integrated GIS environment, Journal of Transportation Engineering 119, pp. 239–254 (1993).
109. Lin G.H., M. Fukushima, New relaxation method for mathematical programs with complementarity constraints, Journal of Optimization and its Applications 118, pp. 81–116 (2003).
110. Lin, G.H., M. Fukushima, A modified relaxation scheme for mathematical programs with complementarity constraints, Annals of Operations Research 133, pp. 63–84 (2005).
111. Lindner D., L.R. Batta, M. Karwan, Equitable sequencing of a given set of hazardous materials shipments, Transportation Science 25, pp. 124–137 (1990).
112. List G.F., P.B. Mirchandani, M.A. Turnquist, K.G. Zografos, Modeling and analysis for hazardous materials transportation: Risk analysis, routing/scheduling and facility location, Transportation Science 2, pp. 100–114 (1991).
113. Liu C., T.C. Chang, L.F. Huang, Multi-objective heuristics for the vehicle routing problem, International Journal of Operations Research 3 (3), pp. 173–181 (2006).
114. Lombard K., R.L. Church, The gateway shortest path problem: Generating alternative routes for a corridor location problem, Geographical Systems 1, pp. 25–45 (1993).
115. Loucks J.S., F.R. Jacobs, Tour scheduling and task assignment of a heterogeneous workforce: a heuristic approach, Decision Sciences 22 (4), pp. 719-738 (1991).
116. Luo Z.Q., J.S. Pang, D. Ralph, Mathematical programs with equilibrium constraints, Cambridge University Press, Cambridge (1996).
117. Lutton J.L., E. Bonomi, Simulated annealing algorithm for the minimum weighted perfect euclidean matching problem, RAIRO Recherche Operationelle 20, pp. 177–197 (1986).
118. McCord M., A. Leu, Sensitivity of optimal hazmat routes to limited preference specification, INFOR 33, pp. 68–83 (1995).
119. Macharis C., Y.M. Bontekoning, Opportunities for OR in intermodal freight transport research: A review, European Journal of Operational Research 153, pp. 400–416 (2004).
120. Madsen O.B.G., A. Larson, M.M. Solomon, Dynamic vehicle routing systems: Survey and classification, Proceeding of the Tristan IV Conference (2007).
121. Madsen O.B.G., The vehicle routing problem with time windows survey and recent developments, C.T.T. Centre for Traffic and Transport Technical University of Denmark, build 115 DK 2800 Kgs. Lyngby, Denmark (2005).
122. Manthley B., Ram L.S., Approximation algorithms for multi-criteria traveling salesman problems, Proceedings of the 4th Workshop on Approximation and Online Algorithms, Springer-Verlag, Berlin, Heidelberg (2006).
123. Marcotte, P., W. Shiquan, Y. Chen, A cutting-plane algorithm for the linear bilevel programming problem, CRT Report 925 (1993).
124. Marcotte P., G. Savard, Bilevel programming: A combinatorial perspective, in Avis D. et al. (eds), Graph Theory and Combinatorial Optimization. Kluwer Academic Publishers, Boston (2005).
125. Martins E.Q.V., On a special class of bicriterion path problems, European Journal of Operational Research 17, pp. 85–94 (1984).

126. Martins E.Q.V., On a multicriteria shortest path problem, European Journal of Operational Research 16, pp. 236–245 (1984).
127. Martins E.Q.V., J.C.N. Climaco, On the determination of the nondominated paths in a multiobjective network problem, Mathematical Methods of Operations Research 40, pp. 255–258 (1981).
128. Martins E.Q.V., J.L.E. Santos, The labeling algorithm for the multiobjective shortest path problem, (downloadable from website http://www.mat.uc.pt/ẽqvm/cientificos) (1999).
129. Miettinen K., On the methodology of multiobjective optimization with applications, Ph.D. thesis, University of Jyvaskyla, Department of Mathematics, Jyvaskyla, Finland (1994).
130. Miller-Hooks E., H.S. Mahamassani, Optimal routing of hazardous substances in time-varying, Stochastic Transportation Networks, Dept. of Civil Engeenering, University of Texas at Austin (1998).
131. Morgenstern C.A., H.D. Shapiro, Chromatic number approximation using simulated annealing, Department of Computer Science, University of New Mexico, Albuquerque, Technical Report (1986).
132. Mote J., I. Murthy, D.L. Olson, A parametric approach to solving bicriterion shortest path problems, European Journal of Operational Research 53, pp. 81–92 (1991).
133. Murata T., H. Ishibuchi, H. Tanaka, Multi-objective genetic algorithm and its application to flowshop scheduling, Computers and Industrial Engineering 30 (4), pp. 957–968 (1996).
134. Murata T., R. Itai, Multi-objective vehicle routing problems using two-fold EMO algorithms to enhance solution similarity on non-dominated solutions, Lecture Notes in Computer Science 3410, pp. 885–896 Springer-Verlag, Berlin, Heidelberg (2005).
135. Nahar S., S. Sahni, E. Shragowitz, Simulated annealing and combinatorial optimization, Proceedings of the 23rd Des. Automation Conference, Las Vegas, pp. 293–299 (1986).
136. Narasimham R., An algorithm for single shift scheduling of hierarchical workforce, European Journal of Operational Research 96, pp. 113–121 (1997).
137. Nguyen S., S. Pallottino, M. Gendreau, Implicit enumeration of hyperpaths in logit models for transit networks, Transportation Science 32, pp. 54–64 (1998).
138. Ombuki B., B.J. Ross, F. Hanshar, Multi-objective genetic algorithms for vehicle routing problem with time windows, Applied Intelligence 24 (1), pp. 17–30 (2006).
139. Outrata J.V., M. Kocvara, J. Zowe, Nonsmooth Approach to optimization problems with equilibrium constraints, Kluwer Academic, Dordrecht (1998).
140. Paixão J.M.P., E.Q.V. Martins, M.S. Rosa, J.L.E. Santos, The determination of the path with minimum-cost norm value, Networks 41 (4), pp. 184–196 (2003).
141. Papadimitriou C.H., M. Yannakakis, On the approximability of trade-offs and optimal access of web sources, in Proceedings of the 41st Annul IEEE Symposium on Foundations of Computer Science, pp. 86-92, IEEE Computer Society (2000).
142. Patel M.H., A.J. Horowitz, Optimal routing of hazardous materials considering risk of spill, Transportation Research A 28 (2), pp. 119–132 (1994).
143. Pirlot M., General local search methods, European Journal of Operational Research 92 (3), pp. 493–511 (1996).

144. Raupach M.R., G. Marland, P. Ciais, C. Le Quéré, J.G. Canadell, G. Klepper, C.B. Field, Global and regional drivers of accelerating CO_2 emissions, Proceedings of the National Academic Science 104 (24), pp. 10288–10293 (2007).
145. ReVelle D.J., C. Cohon, J. Shobrys, Simultaneous siting and routing in the disposal of hazardous wastes, Transportation Science 25, pp. 138–145 (1991).
146. Ruzika S., Wiecek M., Approximation methods in multiobjective programming, Journal of Optimization Theory and Applications 126 (3), pp. 473–501 (2005).
147. Saccomanno F., A. Chan, Economic evaluation of routing strategies for hazardous road shipments, Transportation Research Record 1020, pp. 12–18 (1985).
148. Sancho N.G.F., A new type of multiobjective routing problem, Engineering Optimization 14, pp. 115–119 (1988).
149. Schaffer J.D., Multiple objective optimization with vector evaluated genetic algorithms, in: Grefenstette J.J. (ed.), Genetic Algorithms and Their Applications: Proceedings of the Third International Conference on Genetic Algorithms, pp. 93–100 Lawrence Erlbaum, Hillsdale, NJ (1985).
150. Scheuerer S., A tabu search heuristic for the truck and trailer routing problem, Computers and Operations Research 33, pp. 894–909 (2006).
151. Scholtes S., M. Stohr, Exact penalization of mathematical programs with equilibrium constraints, SIAM Journal on Control and Optimization 37, pp. 617–652 (1999).
152. Sechen C., A.L. Sangiovanni-Vincentelli, The timber wolkplacement and routing package, IEEE J. Solid State Circuits SC-20, pp. 510–522 (1985).
153. Seo K.S., G.S. Choi, The genetic algorithm based route finding method for alternative paths, IEEE International Conference on Systems, Man, and Cybernetics 3, pp. 2448–2453 (1998).
154. Serafini P., Some considerations about computational complexity for multi objective combinatorial problems, in Jahn J., W. Krabs (eds), Recent advances and historical development of vector optimization, Lecture Notes in Economics and Mathematical Systems 294, pp. 222–232, Springer-Verlag, Berlin (1986).
155. Serafini P., Simulated annealing for multiple objective optimization problems, in volume 1 of the Proceedings of the Tenth International Conference on Multiple Criteria Decision Making, Taipei, pp. 87–96 (1992).
156. Shobrys D., A model for the selection of shipping routes and storage locations for a hazardous substance, Ph.D. dissertation, Johns Hopkins University, Baltimore (1981).
157. Skishim C.C., B.L. Golden, Optimization by simulated annealing: a preliminary computational study for the TSP presented at the N.I.H.E., Summer School on combinatorial optimization, Dublin (1983).
158. Sivakumar R., R. Batta, The variance constrained shortest path problem, Transportation Science 28 (4), pp. 309–316 (1994).
159. Sivakumar R.A., R. Batta, M.H. Karwan, A multiple route conditional risk model for transporting hazardous materials, INFOR 33, pp. 20–33 (1995).
160. Skriver A.J.V., K.A. Andersen, A label correcting approach for solving bicriterion shortest-path problems, Computers and Operations Research 27, pp. 507–524 (2000).
161. Srinivas N., K. Deb, Multiobjective optimization using nondominated sorting in genetic algorithms, Evolutionary Computation 2, pp. 221–248 (1994).

162. Steuer R., Multiple criteria optimization: theory, computation and application, John Wiley, New York (1986).
163. Suppapitnarm A., Parks T., Simulated annealing: an alternative approach to true multiobjective optimization, in: Genetic and Evolutionary Computation Conference, Orlando, Florida, pp. 406–407 (1999).
164. T'Kindt V., J.-C. Billaut, Multicriteria scheduling, theory, models and algorithms, Springer-Verlag, Berlin, Heidelberg (2005).
165. Tamaki H., H. Kita, S. Kobayashi, Multi-objective optimization by genetic algorithms: A review, in Fukuda T., T. Furuhashi (eds), Proceedings of the 1996 International Conference on Evolutionary Computation, pp. 517–522, Nagoya, Japan, IEEE (1996).
166. Tanomaru J., Staff Scheduling by a Genetic Algorithm with Heuristic Operators, IEEE International Conference on Evolutionary Computation 1 (29), pp. 1951–1956 (1995).
167. Tarapata Z., Selected Multicriteria Shortest Path Problems: An Analysis of Complexity, Models and Adapatation of Standard Algorithms, International Journal of Applied Mathematics and Computation Science 17 (2), pp. 269-287 (2007).
168. Toth P., D. Vigo, The Vehicle Routing Problem, SIAM Monographs on Discrete Mathematics and Applications (2002).
169. Trivedi M.T., A mixed-integer goal programming model for nursing service budgeting, Operations Research 29 (5), pp. 1019–1034 (1981).
170. Tsaggouris G., C. Zaroliagis, Improved FPTAS for multiobjective shortest paths with applications, Technical Report TR 2005/07/03, Research Academic Computer Technology Institute, Petras, Greece (2005).
171. Turnquist M., Multiple Objectives, Uncertainty and Routing Decisions for Hazardous Materials Shipments, Proceedings of the 5th International Conference on Computing in Civil and Building Engineering ASCE, New York pp. 357–364 (1993).
172. Tuy H., Concave programming under linear constraints, Soviet Mathematics 5, pp. 1437–1440 (1964).
173. Tuy H., A. Migdalas, P. Värbrand, A global optimization approach for the linear two-level program, Journal of Global Optimization 3, pp. 1–23 (1993).
174. Ulungu E.L., J. Teghem, P. Fortemps, D. Tuyttens, MOSA method: a tool for solving multiobjective combinatorial optimization problems, Journal of Multi-Criteria Decision Analysis 8, pp. 221–236 (1999).
175. van Stackelberg H., The theory of market economy. Oxford University Press, Oxford (1952).
176. Vassilvitskii S., M. Yannakakis, Efficiently computing sufficient trade-off curves, Lecture Notes in Computer Science 3142, pp. 1201-1213 (2004).
177. Vicente L.N., P.H. Calamai, Bilevel and multilevel programming: a bibliography review, Journal of Global Optimization 5, pp. 291–306 (1994).
178. Vicente L.N., G. Savard, J. Júdice, Discrete linear bilevel programming problem, Journal of Optimization Theory and Applications 89, pp. 597–614 (1996).
179. Yu P.L., Cone convexity, cone extreme points and nondominated solutions in decision problems with multi-objectives, Journal of Optimization Theory and Applications 14, pp. 319–377 (1974).

180. Warburton A., Approximation of pareto optima in multiobjective shortest path problem, Operations Research 35, pp. 70–79 (1987).
181. Wen U., S. Hsu, Linear bi-level programming problems - A review, Journal of the Operational Research Society 42, pp. 125–133 (1991).
182. White D., G. Anandalingam, A penalty function approach for solving bilevel linear programs, Journal of Global Optimization 3, pp. 397–419 (1993).
183. Wijeratne A.B., M.A. Turnquist, P.B. Mirchandani, Multiobjective routing of hazardous materials in stochastic networks, European Journal of Operational Research 65, pp. 33–43 (1993).
184. Zhang J., J. Hodgson, E. Erkut, Using GIS to assess the risk of hazardous materials transport in networks, European Journal of Operational Research 121, pp. 316–329 (2000).
185. Zhu K., A. Bos, Agent-based design of intermodal freight transportation systems, NECTAR Conference, Delft (1999).
186. Zitzler E., M. Laumanns, S. Bleuler, A Tutorial on Evolutionary Multiobjective Optimization, Workshop on Multiple Objective Metaheuristics, Springer-Verlag, Berlin (2004).
187. Zografos K.G, C.F. Davis, Multiobjective programming approach for routing hazardous materials, Journal of Transportation Engeneering 115, pp. 661–673 (1989).

Index